Monographs in Electrical and Electronic Engineering

Series editors: P. Hammond, T. J. E. Miller, and S. Yamamura

Monographs in Electrical and Electronic Engineering

21. *Brushless permanent-magnet and reluctance motor-drives* (1989) T. J. E. Miller
22. *Vector control of a.c. machines* (1990) Peter Vas
23. *Brushless servomotors: fundamentals and applications* (1990) Y. Dote and S. Kinoshita
24. *Semiconductor devices, circuits, and systems* (1991) Albrecht Möschwitzer
25. *Electrical machines and drives: a space-vector theory approach* (1992) Peter Vas
26. *Spiral vector theory of a.c. circuits and machines* (1992) Sakae Yamamura
27. *Parameter estimation, condition monitoring, and diagnosis of electrical machines* (1993) Peter Vas
28. *An introduction to ultrasonic motors* (1993) T. Sashida and T. Kenjo
29. *Ultrasonic motors: theory and applications* (1993) S. Ueha and Y. Tomikawa
30. *Linear induction drives* (1993) J. F. Gieras
31. *Switched reluctance motors and their control* (1993) T. J. E. Miller
32. *Numerical modelling of eddy currents* (1993) Andrzej Krawczyk and John A. Tegopoulos
33. *Rectifiers, cycloconverters, and a.c. controllers* (1994) Thomas M. Barton
34. *Stepping motors and their microprocessor controls. Second edition* (1994) T. Kenjo and A. Sugawara
35. *Inverse problems and optimal design in electricity and magnetism* (1995) P. Neittaanmäki, M. Rudnicki, and A. Savini
36. *Electric drives and their control* (1995) Richard M. Crowder
37. *Design of brushless permanent magnet motors* (1955) J. R. Hendershott and T. J. E. Miller
38. *Reluctance synchronous machines and drives* (1995) I. Boldea
39. *Geometry of electromagnetic systems* (1996) D. Baldomir and P. Hammond
40. *Permanent-magnet d.c. linear motors* (1996) Amitava Basak

Permanent-Magnet DC Linear Motors

Amitava Basak
Cardiff School of Engineering
University of Wales College of Cardiff

CLARENDON PRESS · OXFORD
1996

Oxford University Press, Walton Street, Oxford OX2 6DP
Oxford New York
Athens Auckland Bangkok Bombay
Calcutta Cape Town Dar es Salaam Delhi
Florence Hong Kong Istanbul Karachi
Kuala Lumpur Madras Madrid Melbourne
Mexico City Nairobi Paris Singapore
Taipei Tokyo Toronto
and associated companies in
Berlin Ibadan

Oxford is a trade mark of Oxford University Press

Published in the United States by
Oxford University Press Inc., New York

A catalogue record for this book is available from the British Library

Library of Congress Cataloging in Publication Data
(Data applied for)
ISBN 0 19 859392 9

Typeset by the author
Printed in Great Britain by
Bookcraft (Bath) Ltd
Midsomer Norton, Avon

to

Shyamali

PREFACE

No standard work of reference dealing with direct-current linear motors in all their aspects has ever been published. However, a considerable literature in the form of published papers dealing with this subject, and also hitherto unpublished work, particularly of an industrial or applied nature, has accumulated during the last 25 years. An attempt has been made to collate all this information and to present it in a comprehensive and orderly manner in this unique volume.

This book has been designed to be useful to two main categories of reader, namely, electrical and mechanical engineers in the user industries, and post-graduates and students of mechanical and electrical engineering.

The first chapter is a general discussion of linear motors. Although the general public became aware of d.c. linear motors in the 1950s these motors have been around since the last century. Readers will welcome this first chapter because it is difficult to find books that give a general discussion on linear motors as a whole. Direct-current linear motor configurations, which are mainly designed for comparatively short distances, are described in the second chapter. This book deals with d.c. linear motors whose magnetic sources are permanent magnets rather than electromagnets and therefore the third chapter is devoted mainly to the properties and to the production of permanent-magnet materials which are presently used in motors, especially ceramic magnets for their low cost, and rare-earth Nd–Fe–B magnets for their very high maximum energy product. Theories are discussed in Chapter 4 on d.c. linear motors of all configurations. Readers will need the equations in this chapter in order to design a motor with a given specification. Chapter 5 is devoted to the supports and guidance systems which are available for these types of motor. Although the magnetic support for alternating-current linear motors is quite advanced in both research and practice, especially in Germany and Japan, some researchers have also successfully developed contactless d.c. linear motors. Methods of controlling these motors are described in Chapter 6. Novel test rigs which are specially designed for linear drives, measurements of different motor parameters, and the computation of various magnetic fields, which all eventually lead to greatly improved design of motors, are elaborately discussed in Chapter 7. In the final chapter, Chapter 8, applications of d.c. linear motors are thoroughly discussed, with examples of different kinds of motor being given. The Appendices give a list of manufacturers of d.c. linear motors with photographs and a comparison of some of their products; a collection of references can be found at the end of the book.

I am grateful to Dr F. J. Anayi for allowing part of his work to be included in this book. I am also grateful to my past and present students, especially to Aly Flores-Filho and Atsushi Matsumura, who have helped design motors and carry out tests. I wish to thank the editorial panel for their helpful criticism of the form of the book, and for their thorough examination of the manuscripts. Finally, acknowledgements are due to my wife Shyamali, and children Somita, Soumen, and Swagota for painstakingly typing the manuscript.

Cardiff, 1996 Amitava Basak

CONTENTS

PRINCIPAL SYMBOLS

a conductor cross-sectional area

A area

B magnetic flux density

BH energy product

C_{f} fringing flux factor

d diameter of conductor

D demagnetization factor; diameter of armature core; damping force coefficient

δ air-gap length

e induced electromotive force; instantaneous voltage

E energy; induced electromotive force

η efficiency

f frequency; force; mechanical load

F magnetomotive force; mechanical force; frictional coefficient

H magnetic field strength

i instantaneous current

I current

J current density

k armature winding factor

K motor coefficient

K_{A} amplifier coefficient

K_{f} thrust coefficient

K_{P} potentiometer sensitivity

l length

L self-inductance; length

L_f leakage-flux factor

L_m motional inductance coefficient

Λ permeance

m mass

M intrinsic magnetization; mutual inductance; mass

μ permeability

n rotational speed; number of conductors

N number of turns per unit length

ϕ flux

ψ flux linkage

p leakage factor

P number of poles; power; magnetic permeance

$\overline{P}$ magnetic permeance coefficient

q reluctance factor

r resistance

R resistance

$\Re$ reluctance

s laplace operator

S surface

t time, thickness

T Maxwell's stress tensor

τ time constant

v velocity

V volume; voltage; velocity

w width

x displacement

z impedance

Subscripts

a armature

b back

c armature core; coercive value; coil

CE collector–emitter

d developed

DF diode forward

e electrical

E effective

eff effective

f field; friction

g air gap

G air gap

i intrinsic

in input

l leakage; vertical component

L load

m magnet; mechanical; motor; magnetizing

M magnet

o steady-state value; output; open circuit

p pole

r remanence; relative; reference

rec recoil

s synchronous; suppressor

S stray

sat saturation

t total; travel; terminal

u useful

w working point

0 free space

1

INTRODUCTION TO LINEAR MOTORS

A linear motor produces motion in a straight line directly, without the use of a crank or any other mechanism for converting rotary motion to linear motion. The history of linear motors goes back as far as the last decade of the nineteenth century.[21] These machines were practically forgotten for nearly half a century, but there has been a genuine revival of interest in them since the 1950s. Although linear motors dispense with the need for gears, belts, etc., which are necessary to obtain linear motion from rotary motors, the latter are still preferred because of the wide range of speed and thrust that can be obtained with the help of gearing at an economic cost. However, it is envisaged that more and more direct-current (d.c.) linear motors will be used in linear drive applications in the future.

1.1 Types of motor

The vast majority of electrical motors are designed to produce rotary motion. Generally, longitudinal movement is obtained over an appreciable distance by developing friction between a surface and a wheel, which is revolved by a rotary motor, for example in rail locomotives; but linear motion can be obtained directly, using the forces of electromagnetism which have been used to produce rotary motion.

As in the case of rotary machines, there are two basic types of linear motor, depending upon the mode of excitation: alternating current (a.c.) and d.c. The two different types of motor have different advantages; for example, a.c. linear motors can be used for movement over long distances, whereas d.c. linear motors can be used, without much auxiliary equipment, for precise movement over short distances.

Alternating-current linear motors can be subdivided into two main groups, (a) linear induction and (b) linear reluctance motors, and both of these types have been thoroughly investigated during the last 40 years. Direct-current linear motors, on the other hand, have been comparatively neglected. There are three types of basic d.c. linear motor: reluctance, conventional, and recorder.

1.2 Alternating-current linear induction motors

1.2.1 *Principles of operation*

Alternating-current linear induction motors are the best known and most frequently used linear motors, since their secondary member (rotor) is, in its simplest form, only a strip of conductive material which does not require an electrical supply. The principle of operation of the induction type of linear motor can best be understood by considering a rotary induction motor with a squirrel-cage rotor; see Fig. 1.1. The rotor consists of copper or aluminium bars which are embedded in slots in the rotor iron and are short circuited at

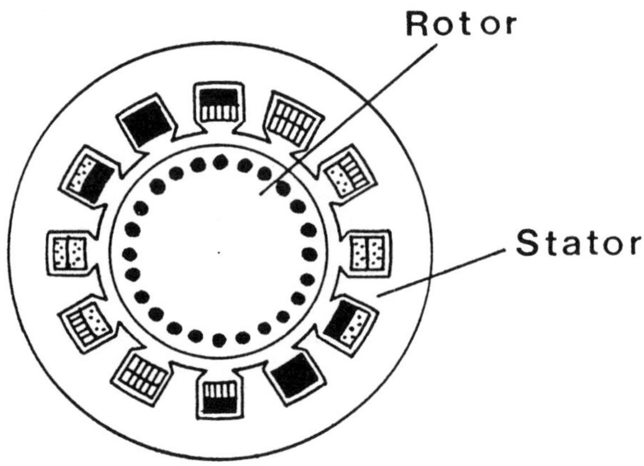

Fig. 1.1. A rotary induction motor with a squirrel-cage rotor.

each end by end rings. The stator windings are wound in such a way that when they are excited from a polyphase supply a travelling magnetic field is produced in the air gap. This travelling field induces eddy currents in the rotor which react with the field to produce rotary motion. This magnetic field rotates at a synchronous speed, n_s, which can be determined from the expression

$$n_s = \frac{120f}{P} , \tag{1.1}$$

where P is the number of poles and f is the source frequency (in hertz). The rotor rotates at a speed which is just less than the synchronous speed, due to losses.

Now, if we cut a squirrel-cage induction motor along a radial plane and then unroll it, a flat configuration is obtained (Fig. 1.2), which is capable of producing motion in a straight line. We can replace the rotor by a thin conducting sheet, but this simplification makes the magnetic circuits of the machine poor. This type of linear motor is known as an open-sided or single-sided machine; see Fig. 1.3(a). If another unrolled stator winding is placed on the other side (the open side) of the single-sided machine, as shown in Fig. 1.3(b), then we obtain a double-sided linear motor.

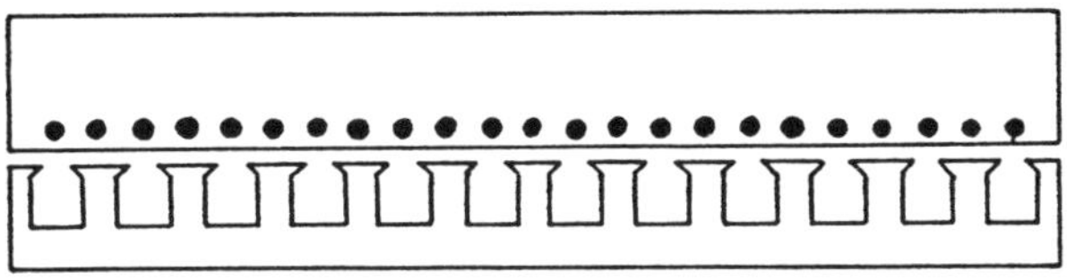

Fig. 1.2. An 'unrolled' induction motor.

Linear induction motors can be classified into two main groups, short-stator (or short primary) and short-rotor (or short secondary) machines, depending on whether the rotor or stator is extended to the full length of the machine. Generally, in the former machines the moving part is the stator and the rotor is extended, whilst in short-rotor machines the moving part is the rotor and the stator is extended to the length of the machine travel.

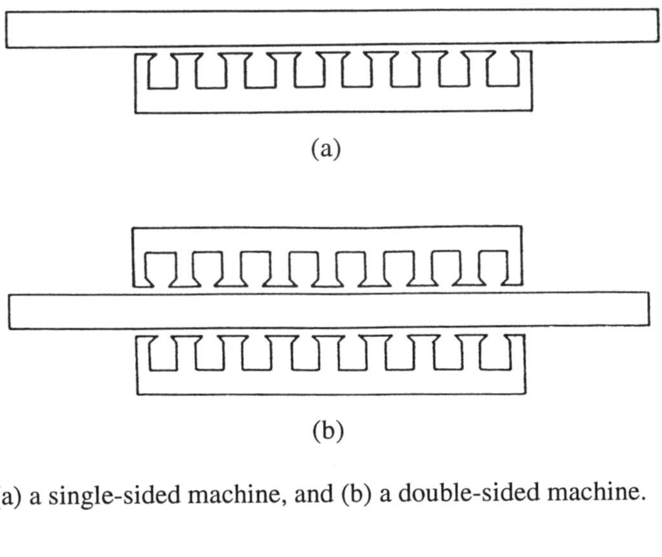

(a)

(b)

Fig. 1.3 (a) a single-sided machine, and (b) a double-sided machine.

Alternating-current linear induction motors have been developed for three particular applications:

(1) oscillating machines, which can be used in the textile industry,
(2) the movement of conductive liquids, and
(3) transportation.

1.2.2 *Oscillating motors*

The earliest references to linear motors is to the induction type,[21] which were proposed as shuttle drives for weaving looms; but, in spite of the invention of many types of linear drive, the use of such motors in the textile industry was not commercially profitable, due to the comparatively low cost of loom mechanics compared with electrical apparatus. This early interest in the use of linear motors as shuttle propellers in weaving looms has been investigated further in the last 40 years.[6,49,50]

In 1957 Laithwaite and Lawrenson[50] reported an investigation into self-oscillating linear induction motors. A short rotor was placed in the influence of two stators, arranged end to end, with their travelling fields directed towards the centre; see Fig. 1.4. The rotor oscillated with a stable amplitude provided that its force–speed curve was in the form shown in Fig. 1.5. Let us assume that the rotor is at rest at position A in Fig. 1.4. During the acceleration period from A to the centre, C, the rotor operates on the portion of the characteristic ap which is shown in Fig. 1.5. It attains a speed v before it crosses the centre line. On entering the reverse field, it is immediately operating at a point q on the force–speed characteristic, and it is brought to rest since it now operates along the portion of the characteristic qa. Since the decelerating force is always less than the accelerating force, the rotor comes to rest at the point B in

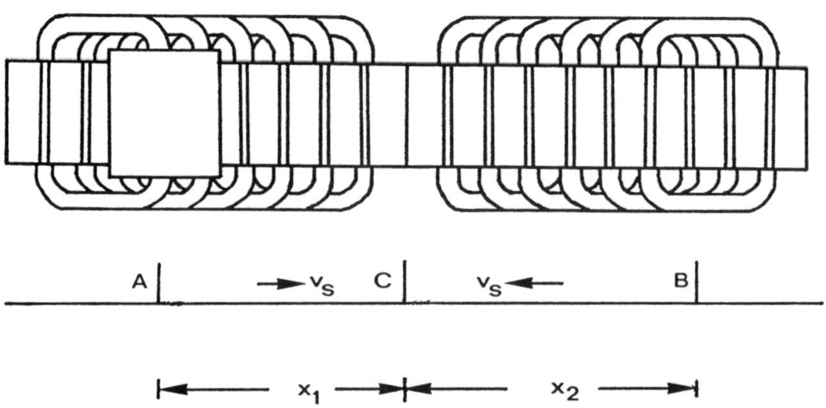

Fig. 1.4. An arrangement of a back-to-back motor for oscillation.

Fig. 1.4, where $x_1 > x_2$. Each successive run produces a rest position which is further away from the centre, until the rotor attains a speed very close to the synchronous speed, v_s, and it oscillates with a constant amplitude. Although this type of motor was first developed for use in weaving looms, it was later suggested that it might be useful in winding packages of yarn.[52] In this application the rotor was required to move the eyelet which guides the thread on to the package. The oscillation of rotors with a mass of order of only 30 g was achieved with speeds of over 500 traverses per minute over a length of 36 cm. The input power was of the order of 100 W.

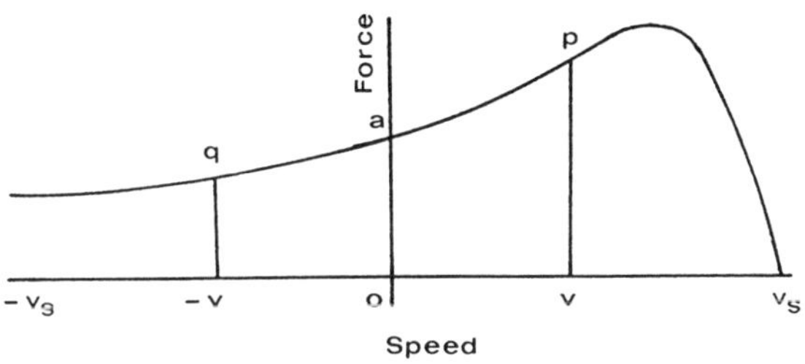

Fig. 1.5. The force–speed characteristic of a self-oscillating induction motor.

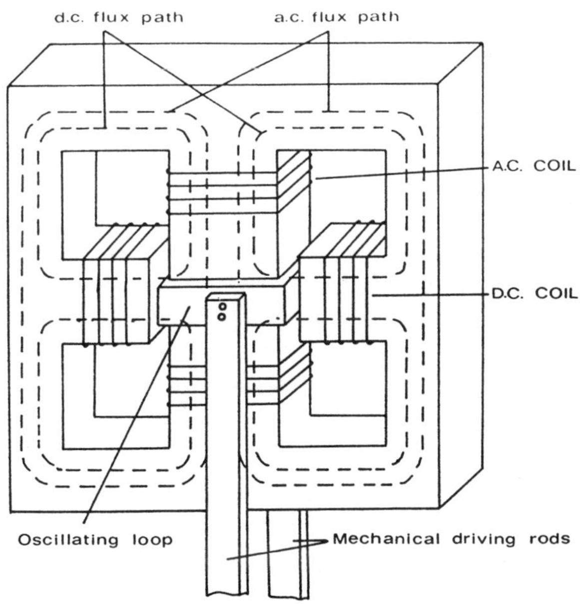

Fig. 1.6. A reciprocating generator.

In 1962, Laithwaite and Mamak[51] developed a new type of a.c. generator. The electromotive force (e.m.f.) induced in this machine appears in the moving member, which consists of a single loop of conducting material, and travels between a d.c. pole structure (Fig. 1.6) along the middle limb with a reciprocating motion. This limb carries stationary coils which experience an induced alternating e.m.f. by transformer action from the moving loop. This typical machine can be used as a generator to convert mechanical power directly from a piston. It can also be used as a motor to drive compressors directly.

An arrangement to overcome the difficulties of making a small oscillating machine was developed in 1962, by West and Jayawant.[81] The machine consisted of two coils which were spaced symmetrically on a stack of electrical-steel laminations. A conducting metal (aluminium, brass, or copper) ring was placed in the space in between the coils. The two coils, which had the same inductance, were connected in series with capacitors whose capacitance was such that the circuit of each coil was nearly resonant. These two series circuits were then connected in parallel to a source with a single-phase supply

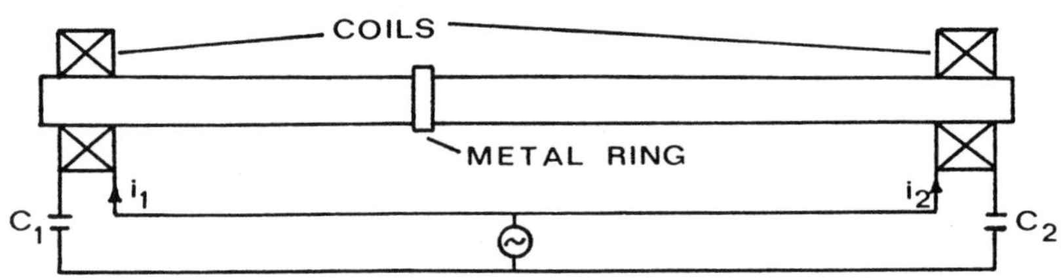

Fig. 1.7. A circuit of a two-coil motor.

(Fig. 1.7). The metal ring oscillated between the coils when the circuit was energized. It was a low power (100 V A) device with an amplitude range of the order of 5–50 cm. Another new linear oscillating machine[63] was described in 1974. This consisted of three parts: a movable part at the centre, carrying a d.c. excitation system, and two fixed C-shaped sections which were excited from an a.c. source (Fig. 1.8).

1.2.3 *Movement of conductive liquids*

Several different types of linear induction pump, which pumped conductive liquid metal, were reported by Woollen[83] in 1955. One of these, known as the flat linear induction pump (FLIP) has a flat primary winding arranged in line (Fig. 1.9). The liquid metal acts as the secondary member of the device and side-bars of copper fixed to the tube walls perform an analogous function to the end rings in rotary induction motors. Although the flat sides of the tube may be only supported by the magnetic cores, it is possible to design a FLIP to handle liquid-metal pressures of up to at least 200 p.s.i. (1.38 MPa). One of the advantages of this type of pump is that the magnetic cores can be withdrawn for repair without disturbing the liquid-metal circuit. The two major disadvantages of FLIP-type pumps are the loss of efficiency due to the absence of side-bars, and the mechanical weakness of the flat sides of the tube. These disadvantages can be overcome by using another type of pump, known as the annular linear induction pump (ALIP). This pump system has an annular space between the central magnetic core and the outer winding, which takes the form of separate rings spaced along the tube; see Fig. 1.10. The liquid metal flows through the annular space. One serious disadvantage of this type of pump is that the windings cannot be taken off for repair without breaking the liquid-metal circuit.

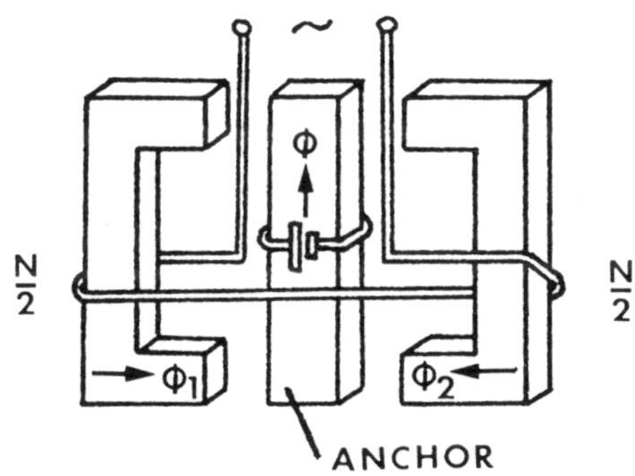

Fig. 1.8. A linear oscillating electric machine.

The principal application of liquid-metal pumps is in movement in nuclear reactors of mixtures of sodium and potassium, which both have high electrical conductivities. The conduction type of electromagnetic pump cannot be used if the liquid metal is appreciably contaminated with nonconductive material; however, the FLIP type of pump has handled sodium–potassium alloys containing considerably more than 0.1 per cent of sodium oxide as the contaminant.

A single-sided linear motor was used in the former USSR in 1962 to pump liquid steel up a 3° sloping channel in a car factory.[22] The liquid steel formed the rotor, and it was separated from the stator (primary winding) by a 6 cm strip of a refractory lining; see Fig. 1.11. The lining could be heated

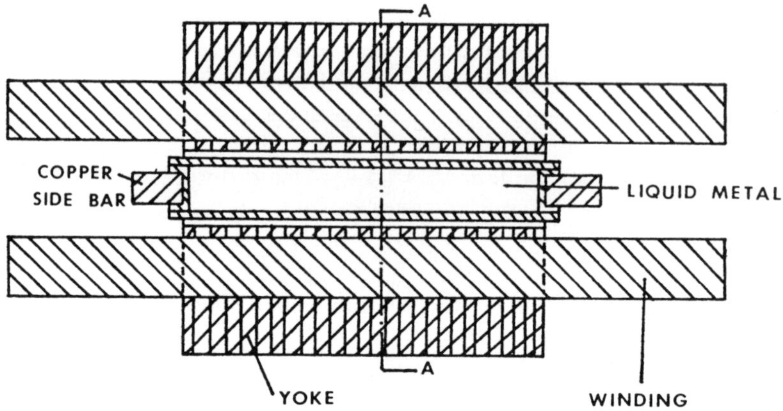

(a)

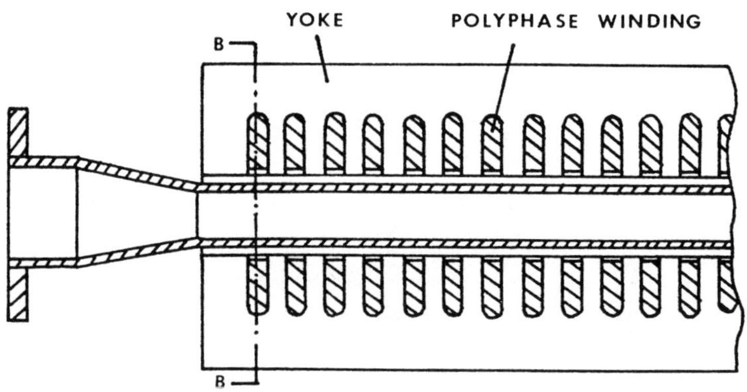

(b)

Fig. 1.9. Cross sections of a flat linear induction pump through: (a) BB, and (b) AA.

before transport of the metal, and in fact the metal itself could be heated during transfer by flame burners in heat-reflecting lids over the channel. When the stator was excited, the molten metal started flowing through the open channel, due to a travelling magnetic field. Since the temperature was well above the Curie point, the liquid steel was nonmagnetic and its resistance was very high; this resulted in the motor having an open magnetic circuit. One of the advantages

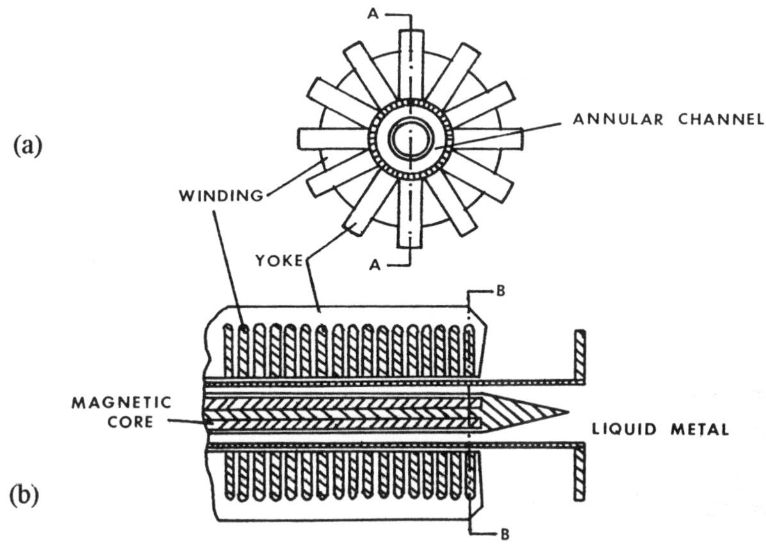

Fig. 1.10. Cross sections of an annular linear induction pump through: (a) BB, and (b) AA.

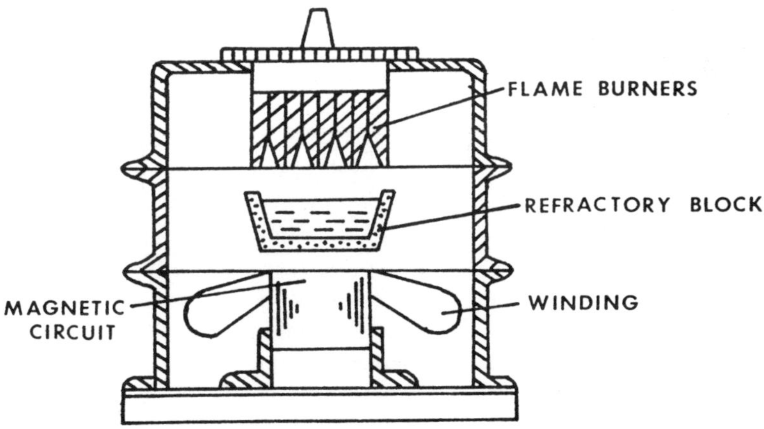

Fig. 1.11. A liquid steel-pump.

of this pump was in the automatic separation of the nonconducting slag from the steel. Although the efficiency of the system was between 0.1 and 1.0 per cent, the system was regarded as a great success because of its many advantages. In 1974, Bamji[7] described a linear machine based on the classical jumping principle, which, despite its open magnetic circuit, low efficiency, and poor power factor, has been successfully used for handling molten aluminium.

1.2.4 *Transportation developments*

In 1905, Wilson[82] proposed the use, for railway traction, of short sections of primary windings (stator) mounted between the rails and the secondary member (rotor) carried on the vehicle. The sections had to be switched in succession to minimize the power requirements. This system was clearly very expensive, since the windings had to extend along the whole length of the track. In the same year, Zehden[85] interchanged the roles of the rotor and stator and placed the primary winding, which is the expensive member, on board the vehicle. The flux path, in this case, was completed through a conducting plate mounted on the track. This was the first moving-primary sheet rotor a.c. linear induction motor.

Throughout transportation history, there has always been a demand for higher speed. In the 1960s, work[48,77] was carried out to develop a ground-transportation system capable of very high speeds, typically 400–550 km.p.h. The US Department of Transportation[77] and Laithwaite and Barwell[48] were among many who considered the suitability of linear motors as high-speed railway drives. A linear motor does not need adhesion for drive and it can be suspended above the ground, to avoid friction, by air-cushion, electromagnet, or permanent-magnet systems.

In 1973, Laithwaite[47] reported the transverse-flux machine, suitable for high-speed transportation. In this machine, the path of the magnetic flux lies in a plane which is transverse to the direction of motion. Figure 1.12 shows two linear axial-flux motors placed side by side; the windings are arranged in such a way that the polarity of one is always laterally opposite to that of the other. Thus the core depth in both the primary and the secondary member is just sufficient to convey flux from one tooth of one motor to the corresponding tooth of the other. Therefore, by turning the magnetic circuit of an axial-flux motor through a right angle, the core depth in both the primary and the secondary member was reduced; and at the same time the standing waves and other undesirable end effects were removed.

An immense amount of analysis and development have been carried out all over the world on the levitation and propulsion of a.c. linear induction motors.[1,5,24,40,78] A comprehensive study and appraisal of this work is beyond the scope of this book; however, it should be mentioned that Thyssen Henschel, the German conglomerate, will soon run the world's first long-distance transport system which is propelled by the force of magnetic attraction (this is shown in Fig. 1.13); it will reach speeds of up to 450 km.p.h between Berlin and Hamburg.[53]

1.3 Alternating-current linear reluctance motors

A form of a.c. linear motor known as a rectilinear screw-thread reluctance motor was described by Gerrard and Paul in 1971.[30] The stator windings of this machine are distributed in 12 equally spaced semiclosed radial slots. A two-start screw thread with a square cross section is cut on the inside surface of the stator bore; see Fig. 1.14. The rotor is a mild-steel cylinder with square-cross-section slots, orthogonal to the displacement axis. The slots of the stator and rotor have the same pitch. The equilibrium position of the rotor, when the machine is energized, will be such that its teeth are in line with those stator threads which coincide with the magnetic poles. Any advance or retardation of the phase of the exciting current rotates the magnetic field accordingly; hence

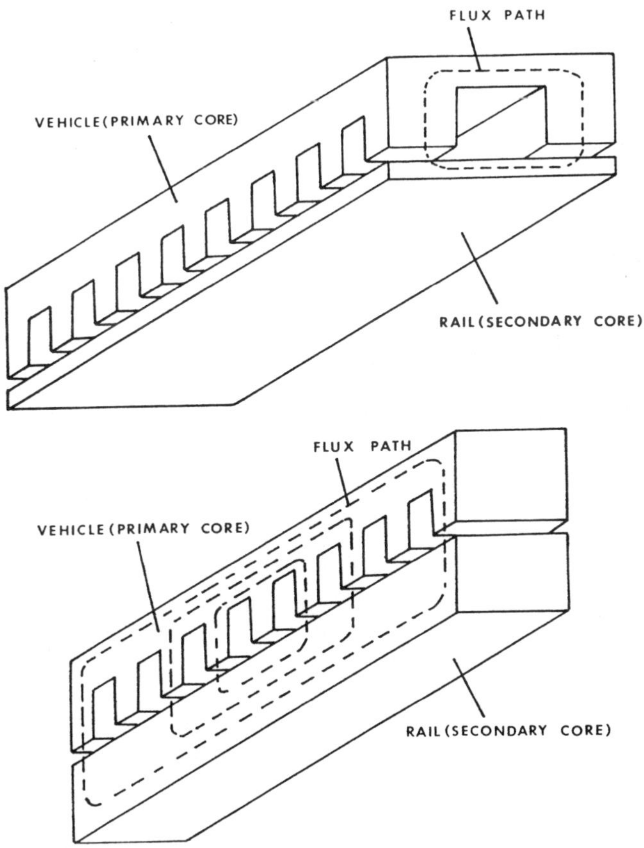

Fig. 1.12. A comparison of a transverse-flux motor and a normal axial-flux single-sided motor showing the flux path and the relative thickness of the motor and the rail.

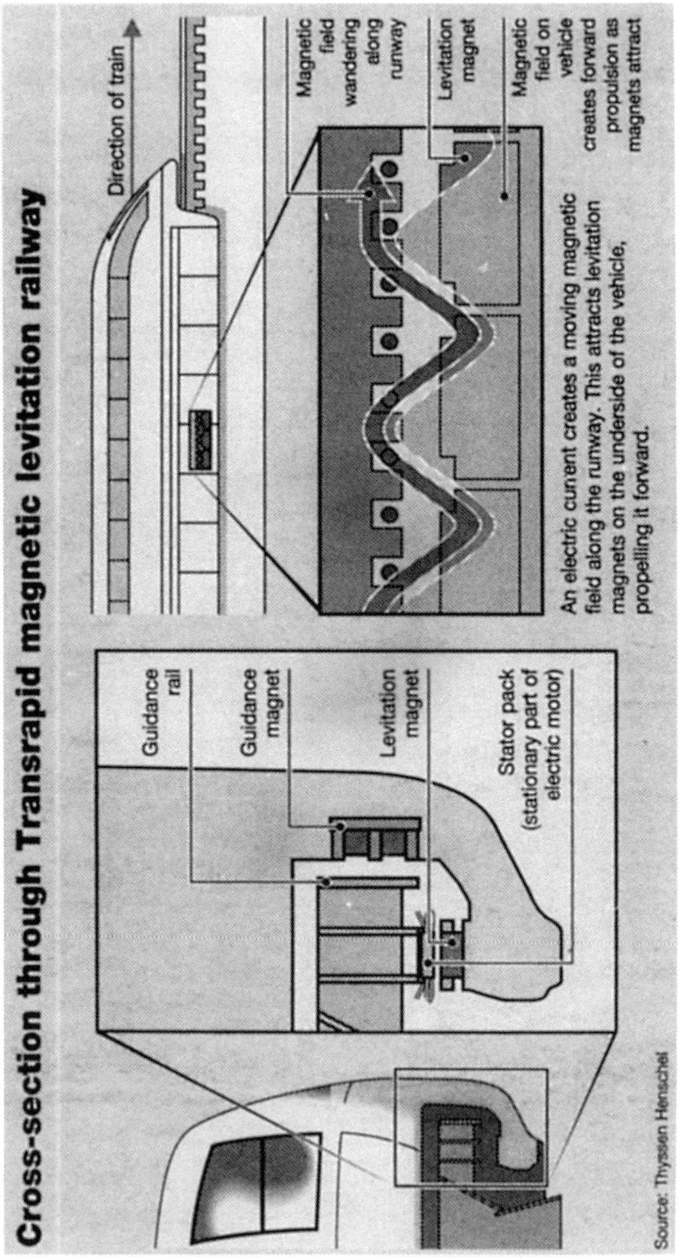

Fig. 1.13. Propulsion by the force of magnetic attraction.

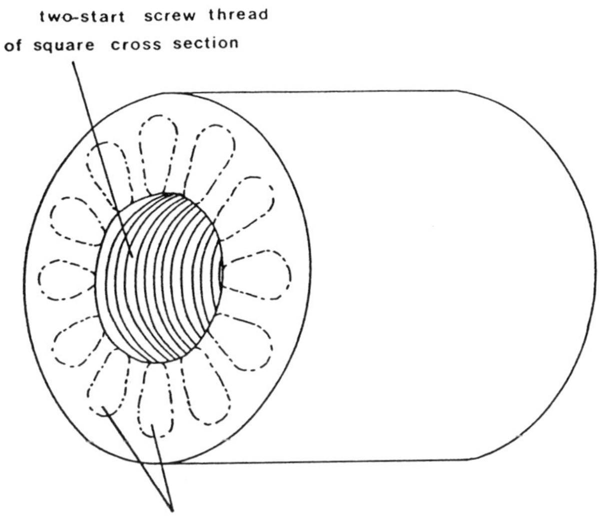

Fig. 1.14. A stator of a rectilinear-screw-thread reluctance motor.

the effect of the screw head is to displace the machine poles along the longitudinal axis of the machine. Thus the rotor will be displaced by two screw threads per cycle of excitation. This motor is capable of operating over displacements of one metre or more with good positional accuracy at relatively low speeds and it may be used as a linear actuator for precise machine-tool operations.

Another form of linear reluctance motor was described by Paul in 1972.[64] This motor may be viewed as an unrolled version of the previous machine. It consists of two flat rectangular plates made of mild steel, which are separated by a small air gap by means of linear bearings. One of the plates, which acts as a stator, is fixed and accomodates a three-phase distributed winding. Small-square cross-section slots are cut on the inside stator surface with a two-start lead. Slots with the same pitch as the stator are cut orthogonally in the moving plate (the armature). Linear displacement is achieved by controlling the phase of the excitation supply in relation to the points of minimum reluctance.

Another new form of motor was developed by Paul and Reid[65] in 1974. The three-phase winding was replaced by 12 stepping coils. An electronic control unit was designed to drive the 12 coils in many different modes. A block diagram of this control unit is shown in Fig. 1.15. This motor can be used as a precise digital actuator.

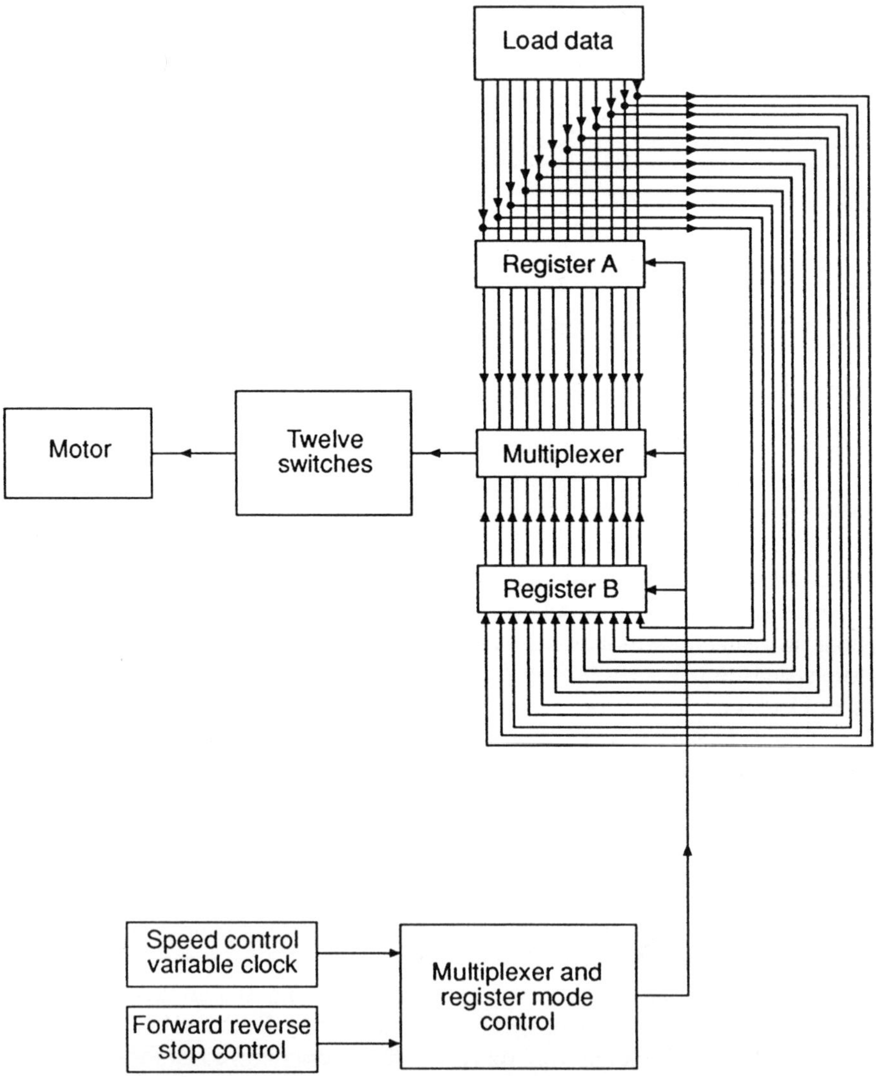

Fig. 1.15. A block diagram of the electronic control unit of a helical reluctance motor.

1.4 Direct-current linear motors

The first d.c. linear motor was built in 1917. It was a reluctance motor in a tubular form; it was proposed as a launcher. A steel missile was attracted into a succession of coils by successively switching the direct currents in the coils; see Fig. 1.16. The motor was known as Birkeland's cannon and it was never developed beyond the model stage.

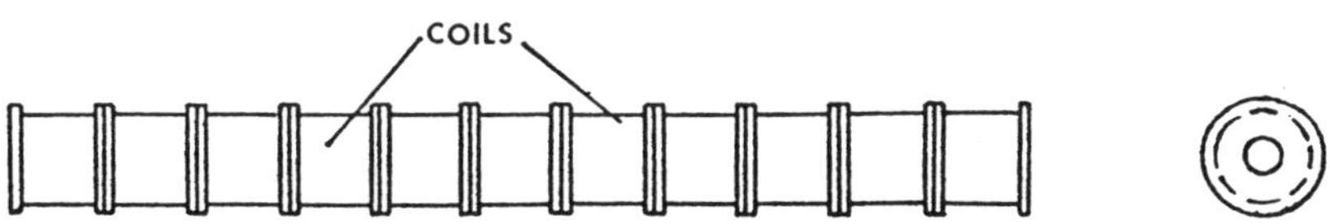

Fig. 1.16. Side and front views of Birkeland's cannon.

Conventional d.c. linear motors are similar to rotating d.c. motors since they have an armature and a field system, but in the linear case the armature is stationary and the field is the moving part. The armature, in general, consists of a circular bar made of a ferromagnetic material, with a single layer of enamelled copper wire closely wound on it. This layer of copper wire acts as both an armature winding and a commutator. The commutator is produced by removing the top surface of the wire insulation along which the brushes run. The field system is mounted inside the carriage, and it consists of a field winding and a shaped yoke. The flux from a pole system crosses the air gap and interacts with the armature current to produce a force, and hence motion.

In 1954, a high-speed d.c. linear motor, a linear form of the conventional rotary motor, was built at the Royal Aircraft Establishment in Farnborough.[72] It was designed as an alternative to a wind tunnel, and it was capable of accelerating a mass of 1 kg to a speed of 500 m s^{-1}. The maximum motor current of 81 000 A was supplied from a very large bank of batteries.

Further experiments were initiated in 1961 as part of the space-research programme in an attempt to simulate meteorite impacts on space capsules.[75] An ingenious d.c. accelerator was built in an attempt to achieve hyper velocities of the order of 15–72 km s^{-1}. Basically, the sliding-coil accelerator employed two coils which attracted or repelled each other. The driving coil, which was stationary, was built by winding an enamelled copper wire on a 25 cm long Bakelite rod 1.9 cm in diameter. The moving coil, A in Fig. 1.17(a), has a

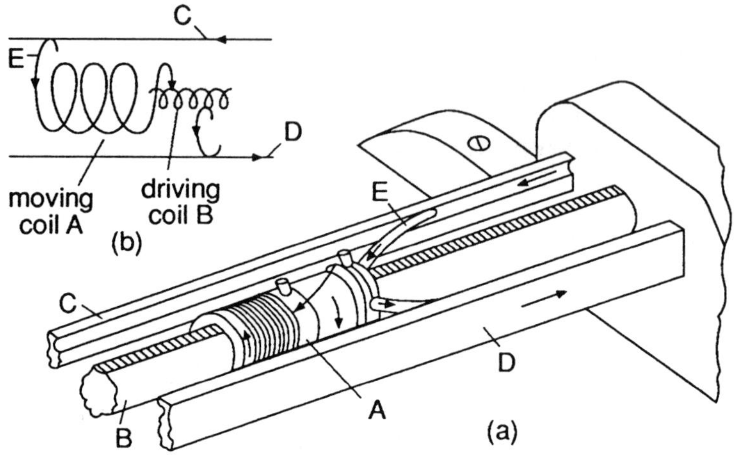

Fig. 1.17. A double-rail accelerator.

diameter which is slightly greater than that of the driving coil B, to allow it to slide freely. The current was supplied through one rail, C, and through a sliding contact, E, to the moving coil. The circuit was completed by the driving coil (Fig. 1.17(b)). A pick-up attached to the moving coil carried the current to the second rail, D. In a different arrangement, shown in Fig. 1.18, only one rail was used, and there were two possibilities of completing the circuit. The current could be fed from the rear end of the driving coil, A, then to the moving coil, B, and through the sliding contact, C, to the rail, D. Here, as the moving

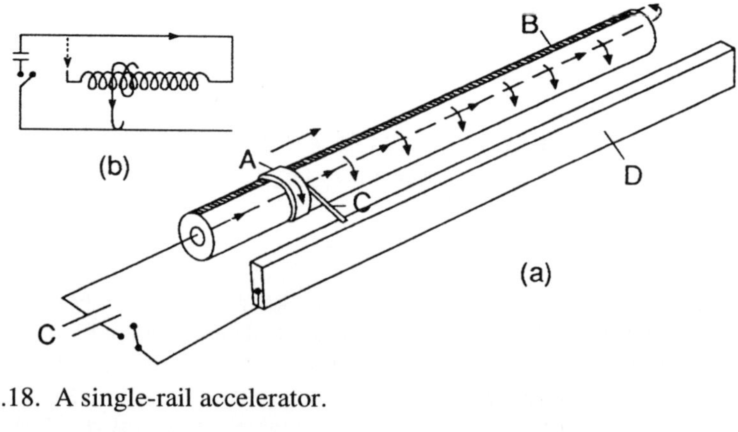

Fig. 1.18. A single-rail accelerator.

coil was pushed along the driving coil, the number of turns to be energized and the ohmic resistance of the circuit were increased. However, when the current was fed into the front end of the driving coil, the moving coil was pulled; thus fewer numbers of turns were energized, and the resistance of the driving coil decreased during the motion of the coil. The best results were obtained when five and ten turn coils were used as projectiles with a driving coil only 10 cm in length. The projectiles acted as a heat sink for the energy which was dissipated; therefore the maximum velocity obtainable without melting is a function of the mass of the projectile. A larger mass is necessary if a higher speed is required, otherwise the projectile will melt before full speed is attained.

Ratcliff and Griffiths[67] developed a novel type of d.c. linear motor in 1964. This machine, like conventional d.c. rotating machines, has a widely adjustable force–speed characteristic, and it can be easily controlled. The track consists of a solid bar of a ferromagnetic material which is overwound with a copper conductor along its entire length (see Fig. 1.19) which acts as an armature. The carriage which surrounds and runs along the armature forms the field circuit; it is made of a ferromagnetic material with an enamelled-copper-wire winding. Brushes fixed to the carriage supply the current to the armature conductors.

In 1969, Green and Paul[31] interchanged the roles of the field unit and the armature of a d.c. linear motor to provide an actuator (see Fig. 1.20) suitable for short-stroke and static applications. The armature, supported by linear

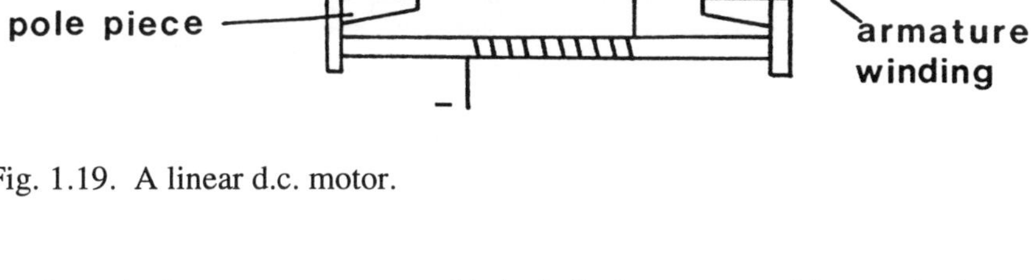

Fig. 1.19. A linear d.c. motor.

bearings, is free to move over a limited displacement and has a single-layer winding, the direction of which is reversed at the midpoint. Thus, neither brushes nor a commutator are required.

Griffiths and Jones[33] also designed a d.c. linear motor in 1969; it had more than two poles. Figure 1.21 illustrates a four-pole machine which has two active poles with field windings and which has two auxiliary poles without windings.

Many chart-recorder manufacturers employ simple d.c. linear motors; they are highly reliable and give a fast speed response from a small unit. One of

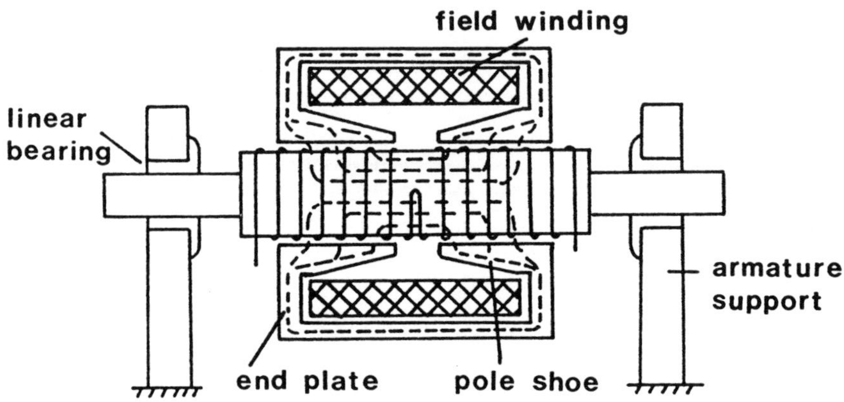

Fig. 1.20. An actuator.

these manufacturers, Foster-Cambridge, has used a d.c. linear motor which has a horizontal soft-iron rod bridging the like poles of two fixed permanent magnets (Fig. 1.22). The instrument's pen and a slide-wire contact are attached to a small coil assembly which moves freely along the rod. The input signal is

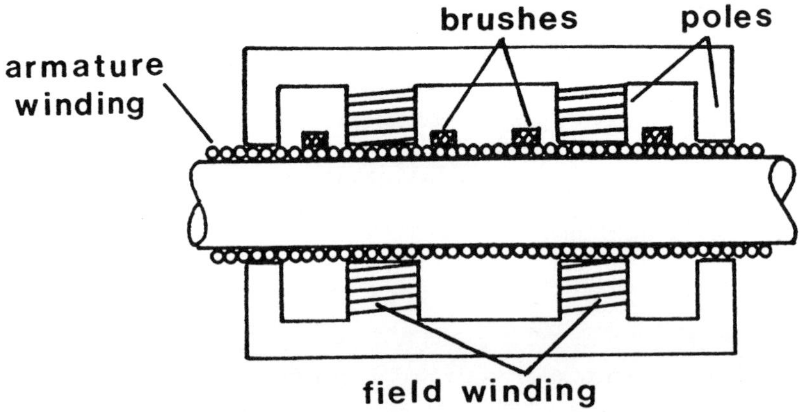

Fig. 1.21. A four-pole d.c. linear motor.

amplified and the amplifier output voltage, V_o is compared with the reference voltage, V_r. This reference voltage is picked off a potentiometer slide wire by the contact attached to the coil; therefore, it depends upon the actual position of the coil with respect to the ends of the potentiometer. The difference in voltage between V_o and V_r is amplified, and the resultant voltage drives the coil. The voltages V_o and V_r become equal when the coil reaches a point on the rod corresponding to the input signal; hence the driving voltage becomes zero, and the coil assembly stops at this point.

Matsui *et al.*[56] developed a d.c. linear motor for high-speed ground transportation. This motor can be controlled by thyristors, and new techniques called *multiplying-coil phase* and *super forced commutation* were introduced to improve the high-speed characteristics. As shown in Fig. 1.23, the magnetic flux of the field magnet on the vehicle interlinks with the vertical and horizontal sides of the coils so that the field magnet receives both a propulsion and a levitation force. The propulsion force is continuously exerted by means of the commutation.

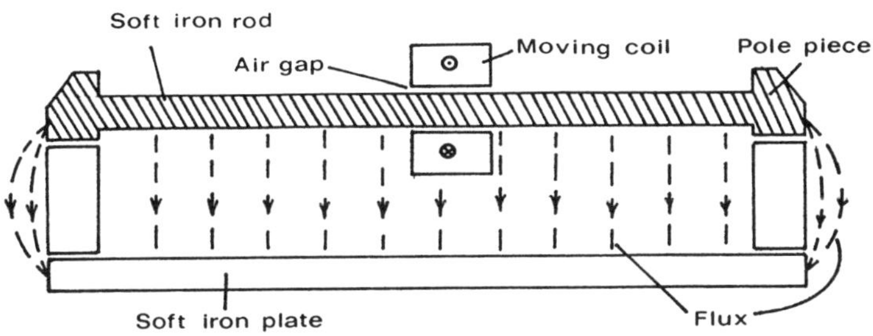

Fig. 1.22. A d.c. linear motor drive used in chart recorders.

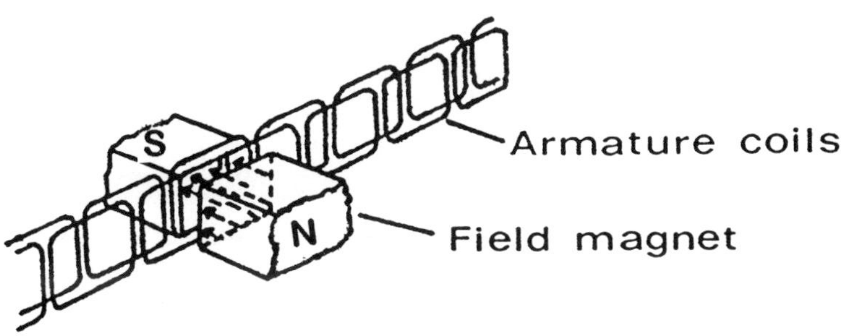

Fig.1.23. A d.c. linear motor used for high-speed ground transportation.

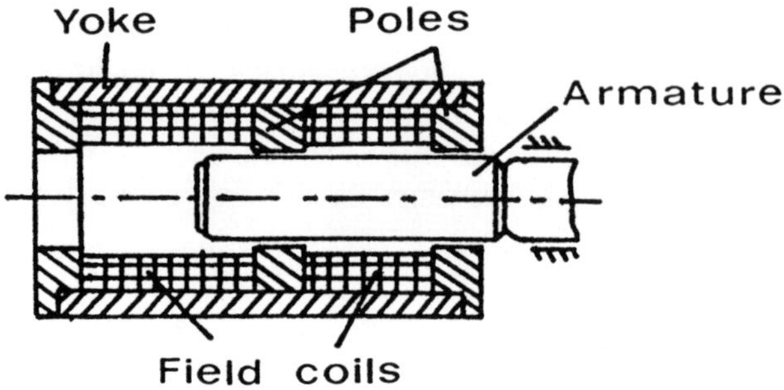

Fig. 1.24. A d.c. linear motor used for impact action.

In 1974 Ryashentev *et al.*[69] designed a motor for impact action. It consists of coils for the forward and reverse strokes, yoke poles, and a travelling armature; see Fig. 1.24. When the coils are fed alternately by voltage pulses the armature performs a reciprocating motion, actuated by the electromagnetic forces of the coils; and thus at the end of a forward stroke, the armature strikes the working tool. The impact pulse is transmitted to an object through this working tool.

This chapter has discussed most typical linear motors, both a.c. and d.c. Each of these motors varies considerably in its operation from the others. According to their principles of operation, d.c. linear motors come under various groups, and they are therefore described in the next chapter in different sections.

2

TYPES OF DC LINEAR MOTOR

Researchers in various countries, such as Japan, Germany, and the UK, have been working on direct-current (d.c.) linear motors for the last 25 years. The basic principles of these motors are similar in essence; but they differ widely in their construction and performance. These two aspects are described in the following sections.

2.1 Configurations of various field units

Some of the factors involved in the design and analysis of two basic forms of d.c. linear motor were presented by Green and Paul.[32] The first type has a multipolepiece field assembly, whereas the second configuration is based on a cylindrical field unit. The former is most suitable when there is a wide range of displacement, and the latter may be used for more restricted ranges of displacement.

2.1.1 *Multipolepiece field units*

A number of mild-steel polepieces, each of which has a surrounding field winding, are placed around the armature at equal intervals; see Fig. 2.1(a). The armature is a mild-steel core with a circular cross section, with a single-layer surface winding, and it is supported at regular intervals by a central web plate which is beneath it. The flux produced by the field unit completes its path via the end plates, pole shoes, air gap, and armature core; see Fig. 2.1(b). The total leakage flux from the field unit forms an appreciable part of the total flux which is established. The measured variation in the leakage flux from the individual members of a two-pole machine with two polepieces is illustrated in Fig. 2.2. The size of the end plate was kept constant, and the lengths of the polepieces were varied. The field excitation was 3000 ampere-turns, and the continuously rated input power was 14 W. For a particular end-plate size there is a polepiece which minimizes the leakage flux.

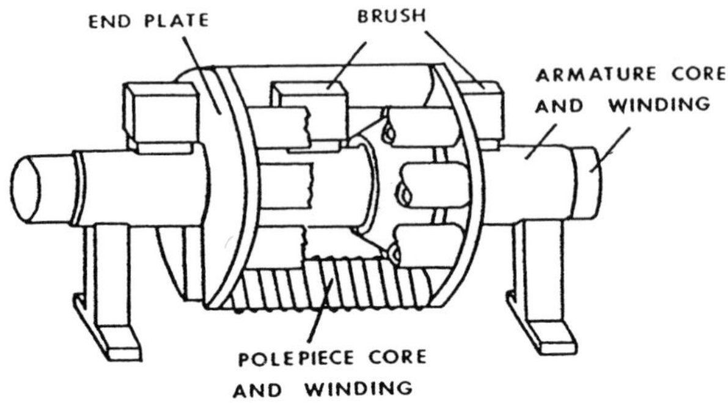

(a)

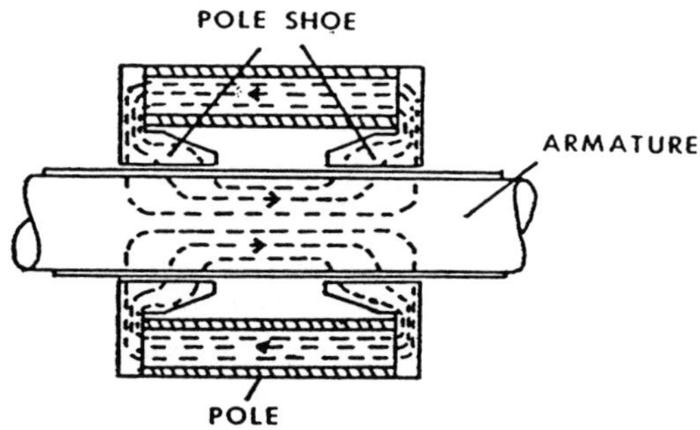

(b)

Fig. 2.1. A multipolepiece field machine: (a) construction, and (b) the flux path.

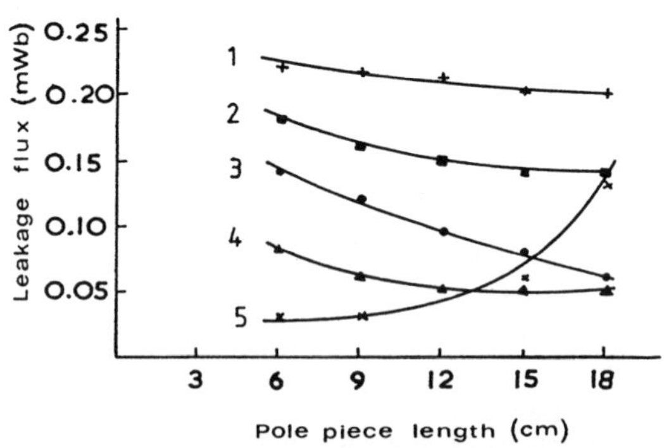

Fig. 2.2. The variation in the leakage flux with the polepiece length: (1) the outer face of the end plate, (2) the edges of the end plate, (3) the inner face of the end plate, (4) the pole shoes, and (5) the pole face.

Again, the leakage of flux from the end plates is dependent on the polepiece length and also on the surface area of the faces. The latter dependence is a direct consequence of the discontinuity in the field unit. The importance of this leakage-flux component could be reduced by increasing the number of poles.

2.1.2 *Cylindrical field units*

The structure of a two-pole linear actuator[31] is shown in Fig. 2.3. The armature, which is supported at each end by a linear bearing and which is free to move over a limited displacement in the longitudinal direction, has a single-layer surface winding. The direction of the winding is reversed at the midpoint, hence neither brushes nor a commutator are required. The field unit encircling the armature consists of a mild-steel outer shell and two end plates which enclose the cylindrical field windings. The main field flux completes its path through the armature core via the air gap, pole shoes, end plates, and outer shell. Unlike multipolepiece field units, where the main field flux is established in the polepieces, the main field flux is established in the armature core in cylindrical machines. Some of this axial flux flows beyond the end plates of the field unit without making any contribution to the force developed. Green and Paul[31] found that, by employing an odd number of poles, this leakage flux could be reduced to about 23 per cent of the total flux. Figure 2.4 illustrates the variation with input power of the useful air-gap flux to the field unit. Both machines have the same number of poles and the same physical dimensions. The multipolepiece unit weighs 1.8 kg whereas the cylindrical unit weighs 2 kg.

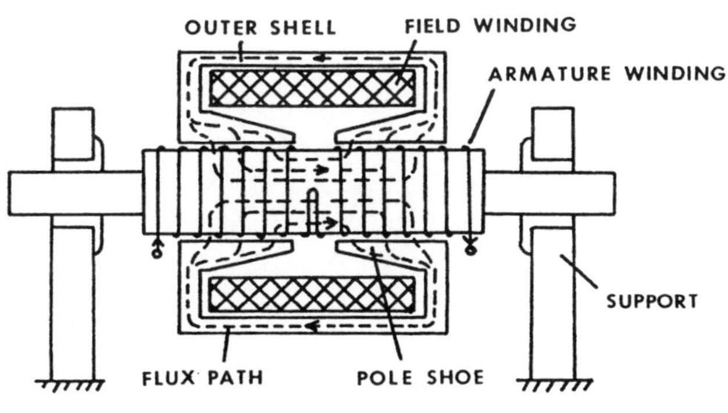

Fig. 2.3. A two-pole linear actuator.

The cylindrical field unit has angular symmetry about the central longitudinal axis of the armature; therefore, the flux density distribution in the regions of differing finite permeability and current density, may be determined analytically.

Static output-force/input-power relationship. An actuator with a specified set of structural parameters produces a thrust which depends on both the armature and on the field excitations. The same thrust can be developed for several different ratios of field and armature excitations. This is confirmed by Fig. 2.5,

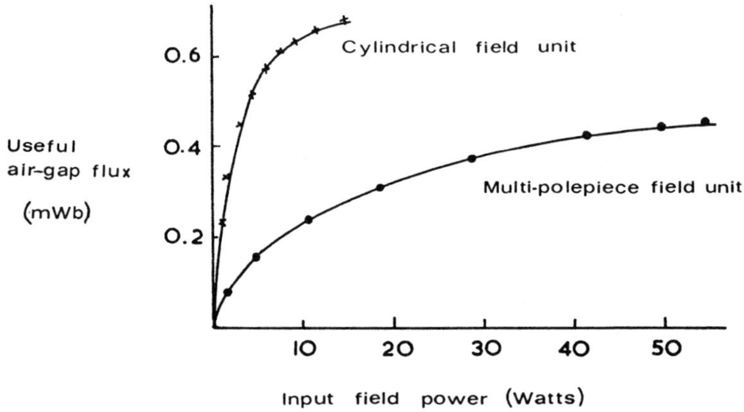

Fig. 2.4. The variation in the useful air-gap flux with the input power for two basic types of d.c. linear motor.

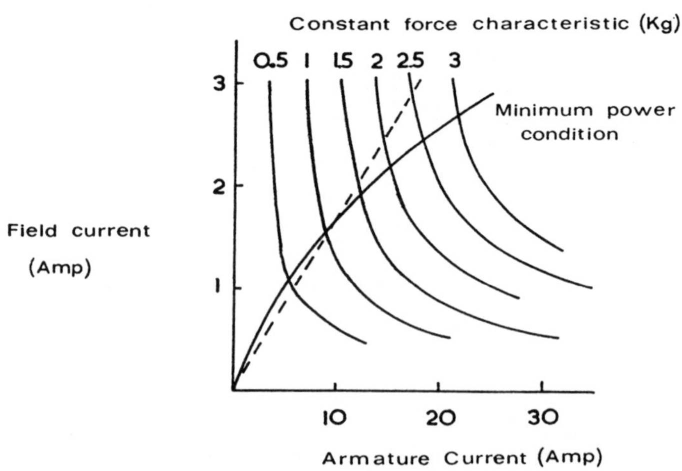

Fig. 2.5. The output force as a function of the armature for six field currents in a two-pole actuator.

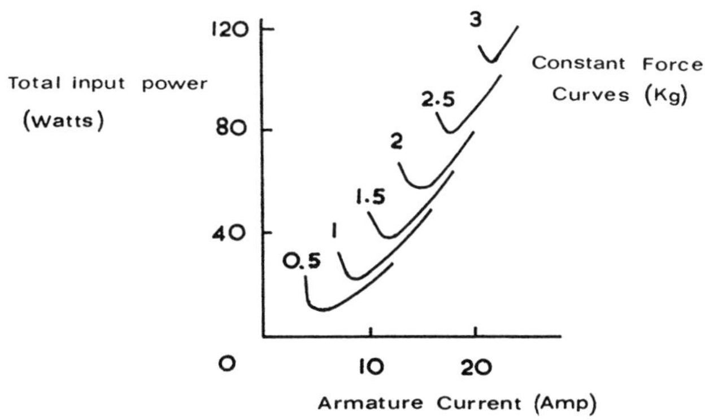

Fig. 2.6. The input-power variation for six constant forces in a two-pole actuator.

which illustrates the variation of the static output force with the armature current for six field currents. The total input power is plotted against the armature current for several output forces in Fig. 2.6. The curve of output force versus total input power shown in Fig. 2.7 can then be obtained by using the minimum-power conditions from Fig. 2.6 and the output-force characteristics of Fig. 2.5.

Force–position relationship. Although cylindrical field actuators have structural symmetry, they are not magnetically symmetrical. Figure 2.8 shows that the flux established by one-half of the armature winding opposes the main field flux, while the flux due to the other half aids the main field flux. Therefore,

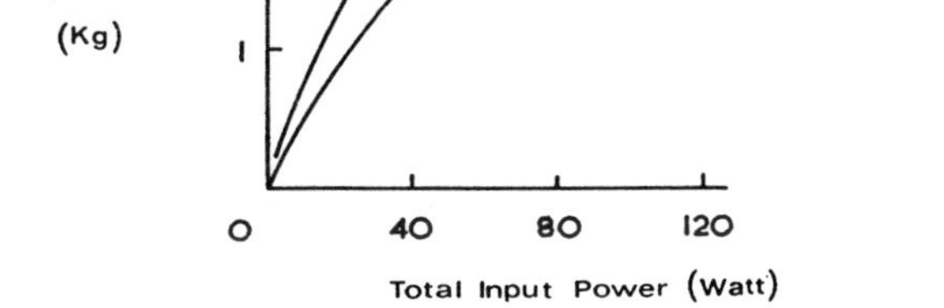

Fig. 2.7. The output-force–input-power relationship.

the flux distribution which results is dependent on the relative positions of the armature and field unit for constant field and armature currents. At the beginning of a stroke, the flux density is low at the armature surface due to a reduction in the net axial flow of flux within the armature core, hence a smaller output force is produced. At the end of the stroke the axial flux is due to the armature and field fluxes, which are in the same direction; this produces a terminal force which exceeds the starting force. Figure 2.9 shows measured and calculated variations in the output force with armature displacement. The output force (and the displacement in the opposite direction) is obtained by reversing either the field or the armature excitation.

(a)

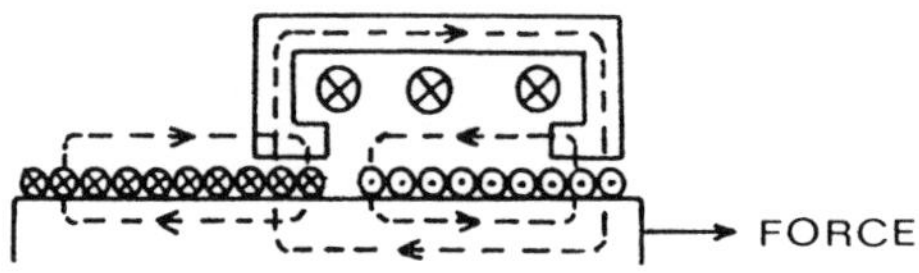

(b)

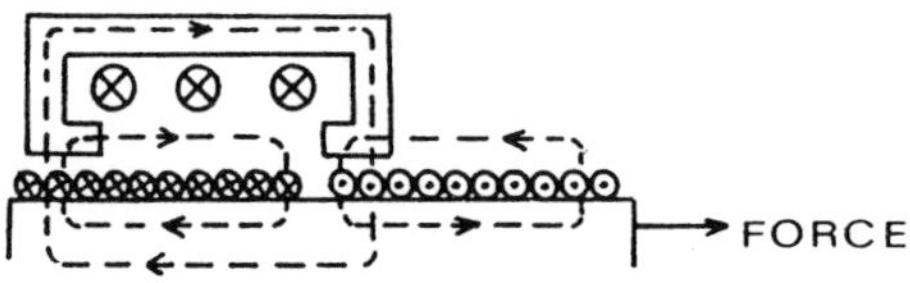

Fig. 2.8. The armature and main-field flux patterns for a two-pole actuator: (a) the armature at the beginning of a stroke, and (b) the armature at the end of a stroke.

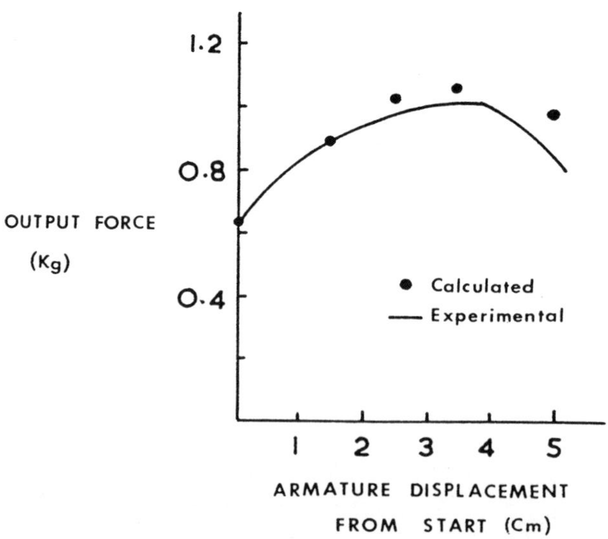

Fig. 2.9. A two-pole actuator with a uniform armature winding.

A graded armature winding permits a variety of force–position characteristics to be obtained, and feedback control can be used in cases where even more flexible characteristics are required. A short-stroke actuator with a graded armature winding could be used in the operation of mechanical valves which require random variations of force with position.[31] The arrangement of a graded armature winding is shown in Fig. 2.10. The output-force–displacement characteristic for a two-pole actuator with a graded winding is shown in Fig. 2.11.

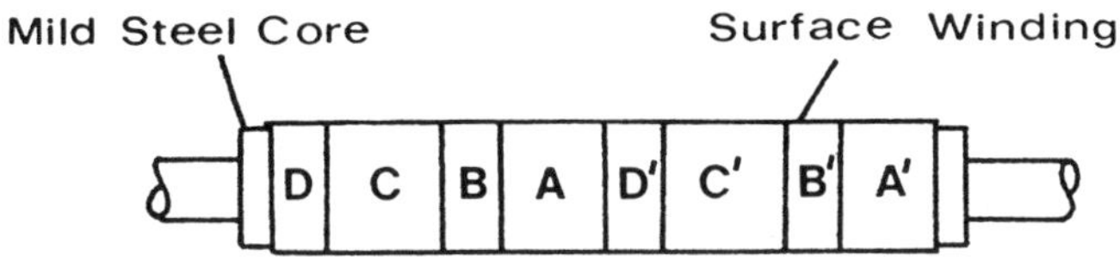

Fig. 2.10. A graded armature winding: ABCD, wound in a clockwise direction; A'B'C'D', wound in an anticlockwise direction; A and A', 22 turns in 19 mm; B and B', 14 turns in 13 mm; C and C', 10 turns in 25 mm; and D and D', 20 turns in 13 mm.

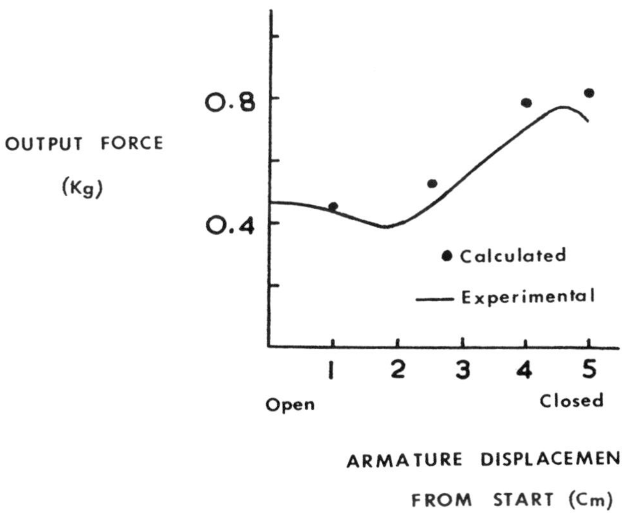

Fig. 2.11. The output force of a two-pole actuator with a graded winding (forward movement).

2.1.3 *Two-pole field units*

A two-pole machine[67] is shown in Fig. 2.12. A mild-steel bar is used for the armature core, which is overwound with a layer of glass tape and on top of this is wound a single layer of enamelled copper wire. Some enamel is taken off the top surface of the winding so it can act as a commutator. The moving part of the machine consists of the flux-generating system and a set of brushes. The flow of current in the armature winding is shown using the general convention, and the flux paths are shown by the dashed lines. The brushes, which are positioned between the poles, are connected as shown in the diagram. The interaction between the current in the armature winding and the flux in the air gap produces a force which enables the field unit to move relative to the armature.

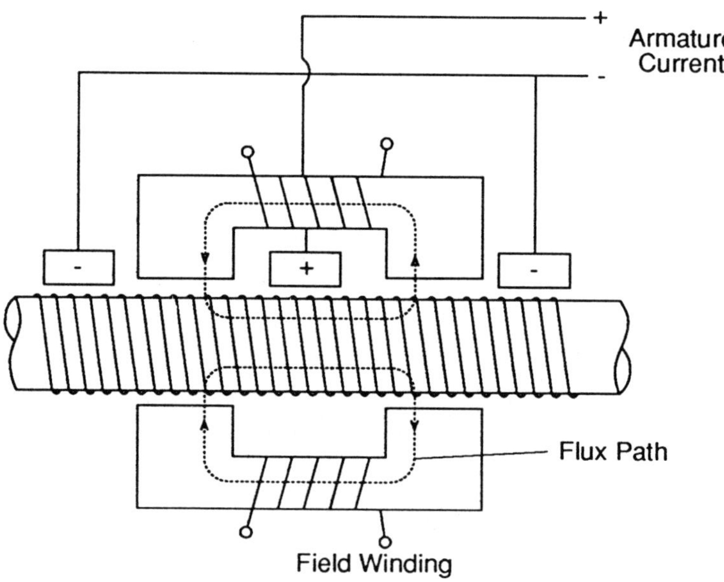

Fig. 2.12. The flux paths in a simple two-pole machine.

2.1.4 *Four-pole field units*

Griffiths and Jones[33] developed a four-pole, separately excited, linear motor with an armature core diameter of 5.1 cm and a winding pitch of 550 turns m^{-1}. The moving unit contained the field source; it was 30 cm long and it weighed about 25 kg. A power input of 5 kW produced a static thrust of 800 N. The no-load speed was almost exactly proportional to the applied voltage, and the maximum speed measured was 9 ms^{-1}. The efficiency of this motor was rather low, only 20 per cent; but, nevertheless, this value was considered to be appreciably greater than that of a comparable alternating-current (a.c.) linear motor. A graph of the speed–load characteristics for a typical four-pole motor is shown in Fig. 2.13. A schematic diagram of such a machine is shown in Fig. 2.14. This configuration is similar to that of a rotating d.c. machine with the field windings around the poles.

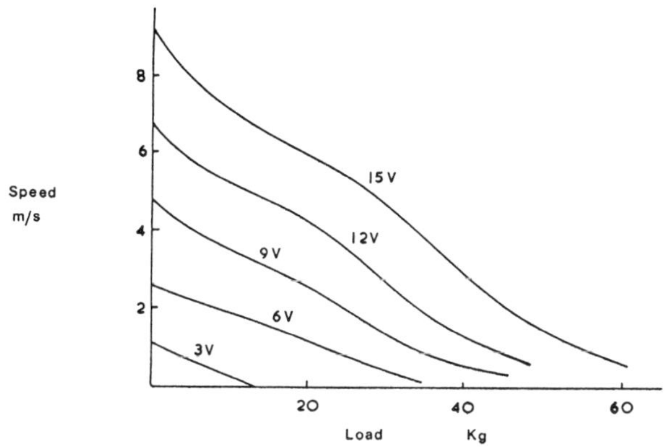

Fig. 2.13. Speed–load characteristics for a four-pole motor.

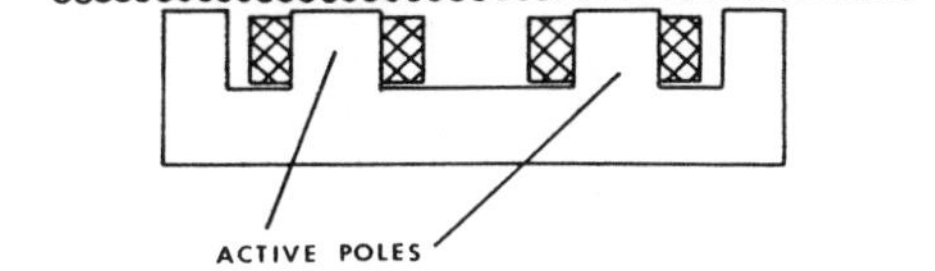

Fig. 2.14. A four-pole machine.

2.1.5 *Six-pole field units*

A schematic diagram of a six-pole machine, with the north and south poles separated on opposite sides of the armature–which reduces the number of field windings–is shown in Fig. 2.15. The flux path is completed via the top plate and it is very much longer than that shown in Fig. 2.14; but, due to the steel top plate, the machine becomes rather heavy. Coils concentric with the armature could be used for the field winding if the armature of the motor were short; see Fig. 2.16.

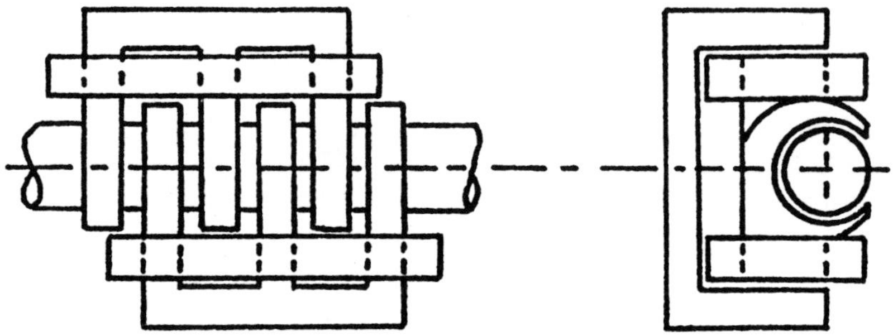

Fig. 2.15. A six-pole machine.

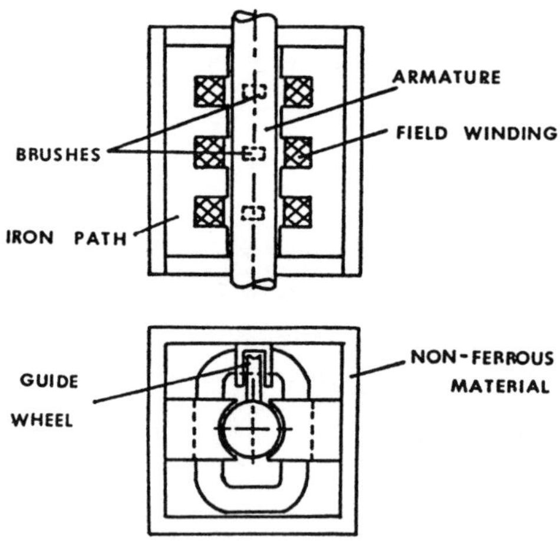

Fig. 2.16. A four-pole machine with concentric coils.

2.2 Brushless direct-current linear motors

While brushless d.c. motors most commonly provide rotary motion, there are several brushless d.c. motor configurations which also provide direct linear motion. These can be moving-magnet, moving-armature, or moving-coil motors. Although there are several differences between these types of motor, they all provide essentially the same basic linear motion. These motors are discussed in the following three subsections.

2.2.1 *Moving-magnet motors*

In their simplest form moving-magnet motors consist of a movable armature assembly containing magnets and a support member which also serves as a magnetic-flux return path. Laminated steel stampings can be used in the stator, with windings inserted much as in rotary-type motors. Figure 2.17 shows a schematic diagram of a moving-magnet motor.[66] This type of motor is useful when high linear force levels are required or where mechanical simplicity is desired. The moving-magnet linear motor shown in Fig. 2.17 is the single-sided type, which has a strong magnetic attraction force between the magnet assembly and the stator. While this force is orthogonal to the linear force direction, it can cause strong bias loading of the linear bearing system. The balanced version has two stators on opposite sides of a moving-magnet assembly, and the magnetic attraction forces therefore cancel out in general.

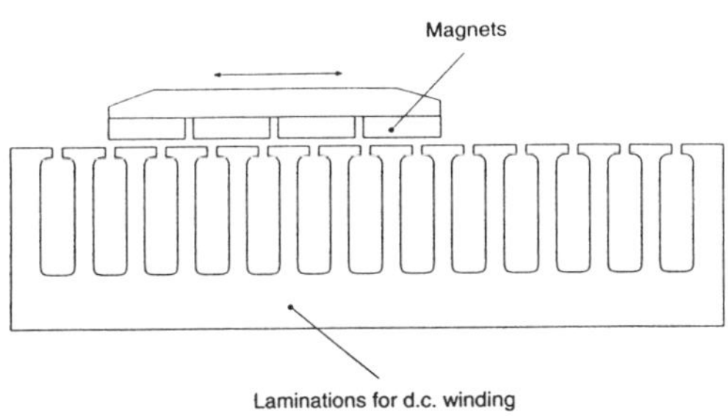

Fig. 2.17. A moving-magnet motor.

2.2.2 *Moving-armature motors*

Moving-armature motors have a moving lamination and winding assembly which is shorter than the stator. The stator consists of a long group of magnets mounted on a magnetically permeable surface. These motors have characteristics which are essentially similar to those in the moving-magnet type, except that all the windings interface with the adjacent magnets. These devices are, therefore, more efficient, but they require more magnets. Also note that moving-armature devices require flexible cables for the windings. Figure 2.18 shows a moving-armature motor.[66]

2.2.3 *Moving-coil motors*

Moving-coil motors,[66] such as that shown in Fig. 2.19, are similar to the moving-armature type, except that the iron is removed from the moving part, and only the conductors and their support structure remain. The magnet structure is of the balanced type, but a structure with magnets on one side and a

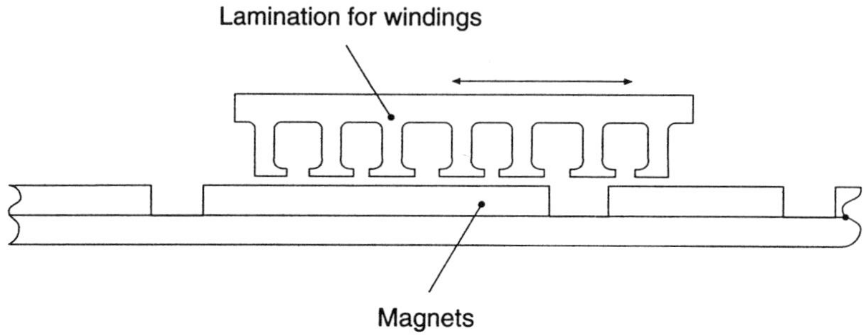

Fig. 2.18. A moving-armature motor.

flux return path on the other side also works. This type of motor, with its light-weight armature, has the highest force-to-inertia ratio. It can be designed to feature force levels with an extremely low force ripple due to the absence of any reluctance forces. Its low moment of inertia allows incremental motion, with extremely high acceleration and deceleration rates. The thermal resistance between the armature and ambient air is higher than in the other models; therefore, the coils may require air cooling at high incrementing rates.

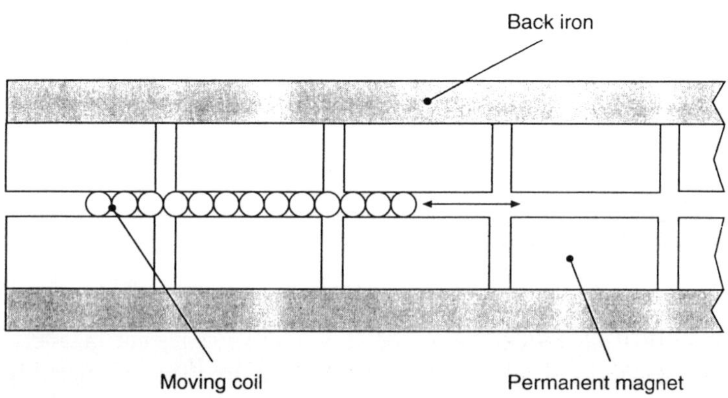

Fig. 2.19. A moving-coil motor.

2.2.4 *A comparison of brushless d.c. motors*

All the types of motor discussed in this section need a positive feedback system for commutation and position-feedback purposes. The most prominent features of the motors are listed in Table 2.1. This comparison should not be taken as absolute; for instance, terms like *low cost* and *high cost* convey the relationship between the costs of two different devices with similar capabilities.

Table 2.1 A comparison of three brushless d.c. linear motors

Linear motor	Moving-magnet	Moving-armature	Moving-coil
Weight of moving unit	Medium	Heavy	Light
End effects	Present	Present	Absent
Power Loss	High	Medium	Low
Moment of inertia	Medium	High	Low
Reluctance clogging	Moderate	Moderate	None
Volume of magnets	Low	Medium	High
Cost	Moderate	Low	High

2.3 Configurations of various stationary armatures

The length of a stationary armature depends on the travel length of the machine. The core material of the armature is mild steel. A glass tape is usually wound on the armature to insulate the core material from the winding. In order to keep the air gap to a minimum, a single-turn winding is commonly used.

2.3.1 *Long- and short-stroke machines*

For long-stroke machines whose length is over 30 cm, supports are necessary for the armature. Guides are also necessary (they are described in Chapter 5). For long travel lengths the moving unit is usually the source of the magnetic field, and therefore it holds the permanent magnets. The travel length can be up to 6 m, and typical speeds of such motors can be up to 3 $m\,s^{-1}$. This type of motor will find applications in materials handling. Figure 2.20 shows a typical example of such a system.[2]

Kano *et al.*[45] developed a long-stroke motor, the structure of which is shown in Fig. 2.21. The stator consists of 24 flat-type air-core coils on a printed circuit board and two rods for guiding the mover. The moving unit has four magnets and a yoke on either side of the flat coils; a rotary encoder and an infrared ray transmitter are carried by the moving unit. A similar motor was described by Persson[66] in 1983.

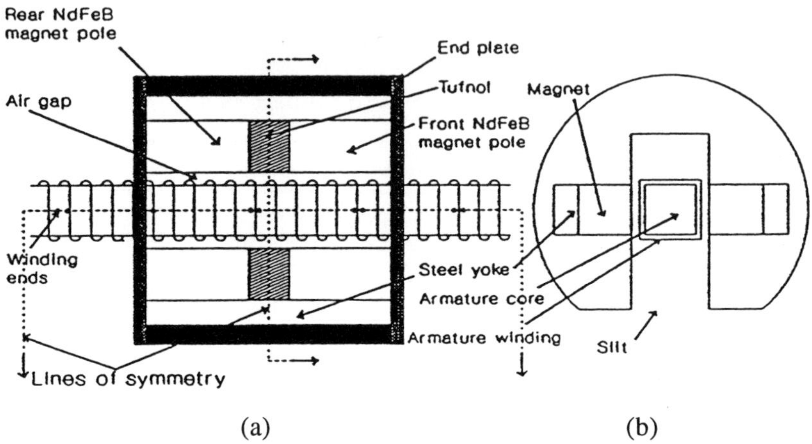

Fig. 2.20. A brushless, 3.5 m long, d.c. linear motor with a square armature: (a) top view, and (b) end view.

For motors with short travel lengths, the moving unit may consist of permanent magnets or it may be a very short armature coil traversing the length of the track. In the first case the moving unit may be built from Tufnol and it may house Nd–Fe–B magnets, and it may travel over the armature (whose windings are in sections), whereas in the second case a coil slides over a mild-steel bar and between two ceramic permanent magnets which are the same length as the track. The ceramic permanent magnets have one-third of the

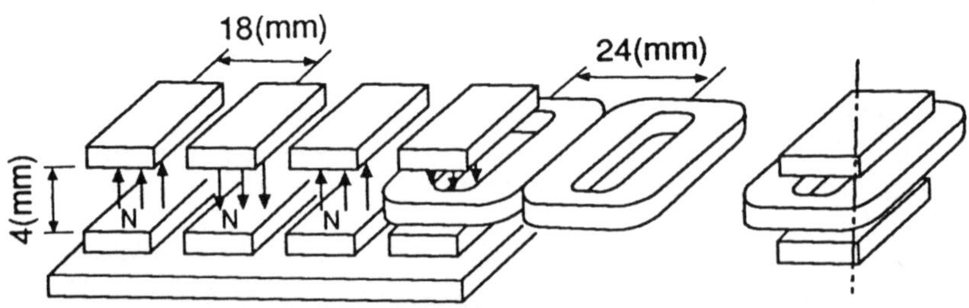

Fig. 2.21. A moving-magnet d.c. linear motor with flat coils.

remanent flux density of the Nd–Fe–B magnets, but they are comparatively much cheaper, as we will see later in this book.

2.3.2 *Circular and rectangular armatures*

Armature cores with circular cross sections are frequently used in motors, but in order to have uniform air gaps arc-shaped magnet poles are also needed. These poles may be expensive to manufacture, and cores of square or rectangular cross section can be used instead. From the manufacturers' point of view it is much easier to wind enamelled wire on a circular core than on a rectangular or square core. The only advantage with rectangular magnet poles is that they will give uniform air gaps, and they are comparatively cheaper than the arc-shaped cores. The motor shown in Fig. 2.20 has an armature with a square cross section with slightly rounded corners. If the diameter of the enamelled copper wire is small, say 1.5 mm, then it will not be difficult to wind, but larger-diameter wire, has to be slowly wound under tension. The windings can be continuous or they can be made of several sections. In the first case, brushes have to be used for long traverses. Movements of motors with the second type of winding have to be sequentially controlled by a set of switching circuits.

2.3.3 *Single and double armatures*

Most motors are designed and built with a single armature located in the middle of the motor. However, researchers have designed motors with two armatures running parallel to each other. The magnetic circuit is completed with two mild-steel blocks placed at the ends of two armatures. From any side of the motor the winding direction is clockwise for one armature and anticlockwise for the other. This doubles the resolution if the same pitch is used for the armature-coil sections. Figure 2.22 shows two types of motor which each have two rectangular armatures.[10,36] In this case, because of the two armatures, the thrust will be doubled for a given current. In a motor described by Basak and Flores-Filho[10] both the armatures have single-layer windings, but it is possible to employ double- or triple-layer windings. A single housing holds all the magnets and this is supported by two linear bearings and by guidance rods.

2.4 Single- and multiple-layer armature windings

The sections of a single layer wound on an armature core can be energized in a sequence, either one at a time or two adjacent cores can be simultaneously energized during an overlap period, in accordance with the speed and position requirements. An overlap period for two consecutive energization times is shown in Fig. 2.23. The software determines the length of the overlap period from a knowledge of the desired and actual positions, and the control circuit energizes and de-energizes the armature sections accordingly. Each of the

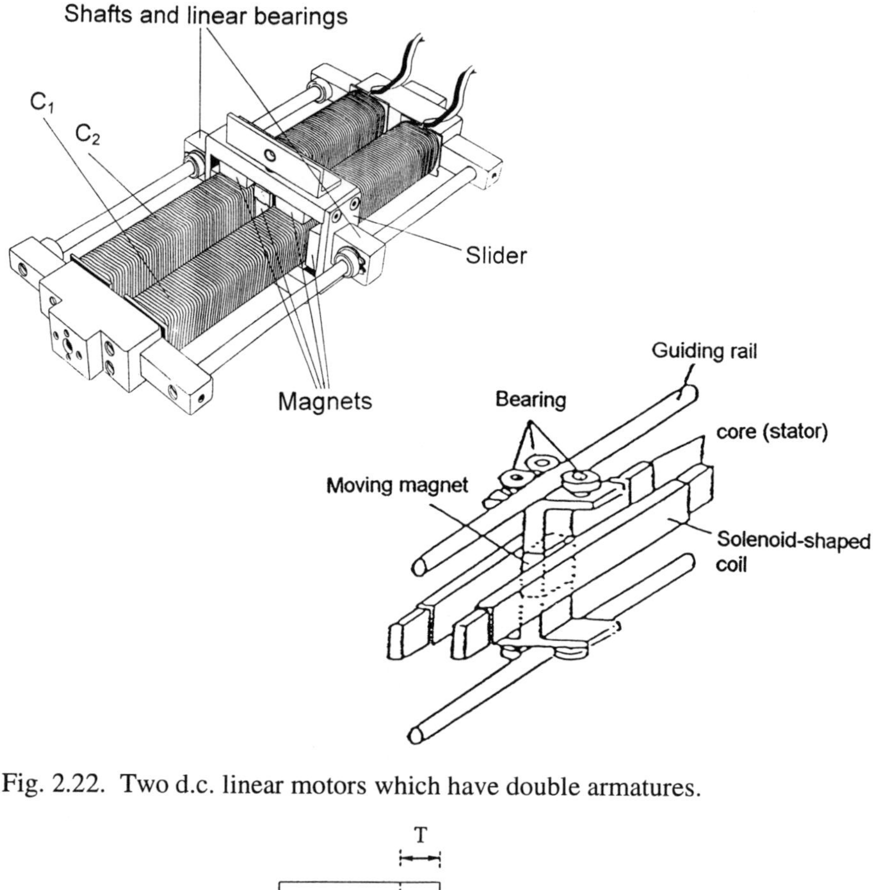

Fig. 2.22. Two d.c. linear motors which have double armatures.

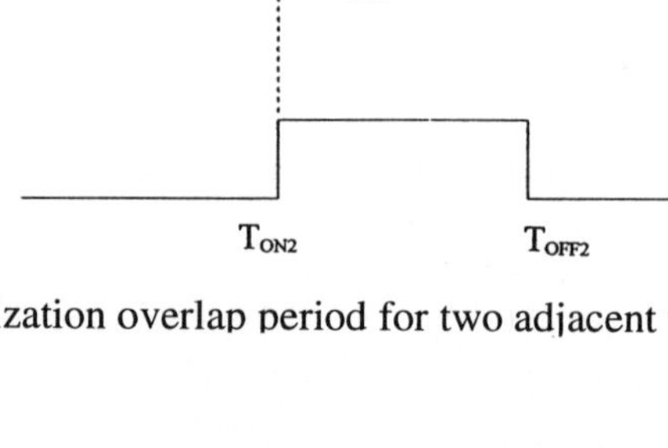

Fig. 2.23. An energization overlap period for two adjacent windings.

armature sections has a switching circuit consisting of a power transistor which is capable of handling up to 30 A. Instead of single-layer armature sections, multilayer armature coils may be used. Since the force developed in a d.c. linear motor is directly proportional to the armature current and to the number of turns per unit length, by increasing the number of turns by a factor of 2 or 3 and at the same time reducing the current by a factor of 1/2 or 1/3, the developed force can be kept to the original value. One way of increasing the number of turns is to have multiple layers of windings. Reduced armature currents demand power supplies with lower ratings, and the power transistors which are required for the switching circuits will require much smaller spaces to be accommodated.

2.5 Special types of motor

Figure 2.24 shows the configuration of a d.c. linear motor which was described by Honds and Meyer[41] in 1982. In this case the static part consists of three soft-iron cores, which can be made of structural steel that has undergone a simple anneal only. The central core is wound directly with insulated copper wire, W; there is no former. The two permanent magnets form the moving part of the motor and they are fixed to a carriage. It is possible, for example, to 'halve' the motor, leaving only one permanent magnet and two iron cores. The magnet can also be stationary, with the coil moving. The current in such a case, however, must be supplied by means of flexible wires or carbon brushes.

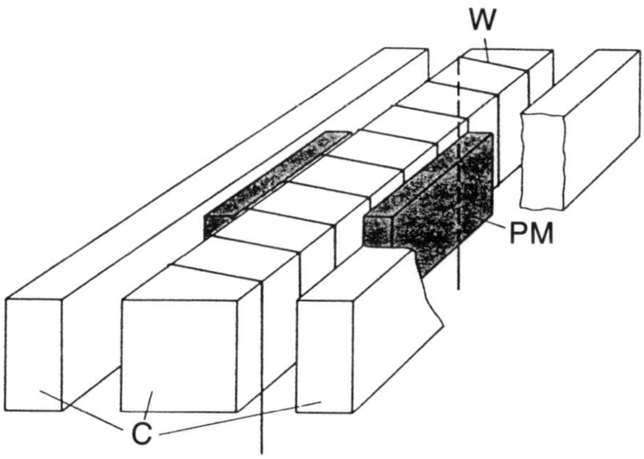

Fig. 2.24. The motor developed by Honds and Meyer.

In 1984 Schultz[71] introduced a d.c. linear motor which can be used to drive a printer head or a recorder pen. The motor, shown in Fig. 2.25, consists of two parts: a fixed rod and moving coils. The rod is surrounded by alternating sections of equally spaced permanent magnets and by nonmagnetic

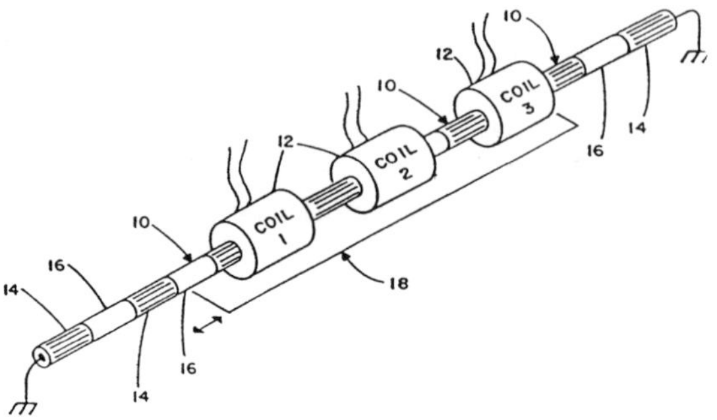

Fig. 2.25. Schultz's motor: (10) nonmagnetic material, (12) coil, (14) permanent magnet, (16) armature rod, and (18) the travel length.

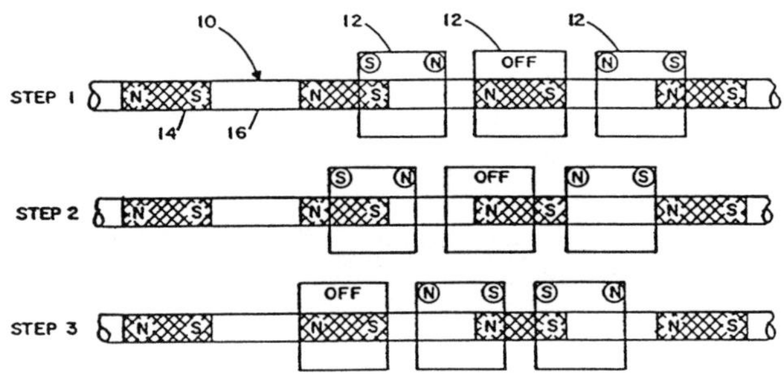

Fig. 2.26. The principle of operation of Schultz's motor.

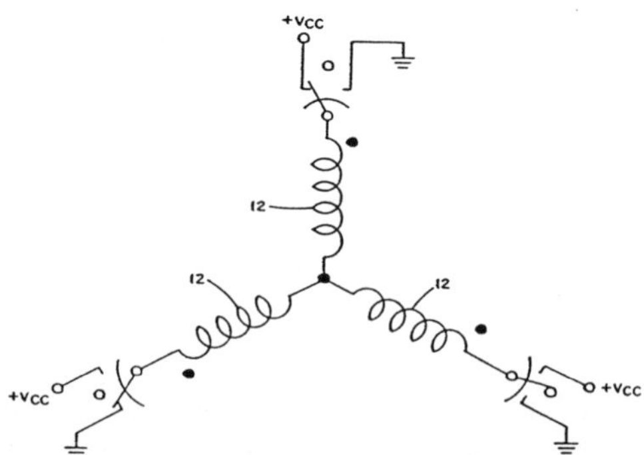

Fig. 2.27. The connection of the phases: (12) coil.

fillers. The motor works by controlling the polarity and by controlling the presence or absence of coil currents in such a way that a continuous motion is produced in the desired direction. Figure 2.26 shows how the motor is propelled from one magnet to the next. The motor needs commutation signals developed by a position-sensing apparatus, and the coils can be connected in a three-phase arrangement, as shown in Fig. 2.27. The speed is controlled by varying the amplitude of the coil currents either by changing the level of V_{CC} or by changing the duty cycle in pulse-width-modulated systems.

Another type of d.c. linear motor, shown in Fig. 2.28, was developed by Belikov *et al.*[20] at the Department of Electrical Machinery of Odessa Polytechnical Institute. It is used in metal-cutting machine tools with wide limits of regulation of the rate of displacement of the moving elements (the minimum velocity is 0.001–0.01 m min^{-1}). The motor has a combined structure, with the armature winding and the excitation winding located at one part of the motor, the moving part, while another part with no windings acts as the shaper of the main magnetic field, and it is provided for this purpose with unexcited pole-shaped projections. The magnetic flux, ϕ_o, is produced by energizing the excitation winding, and the intersection of this field with the armature current produces motion in the motor.

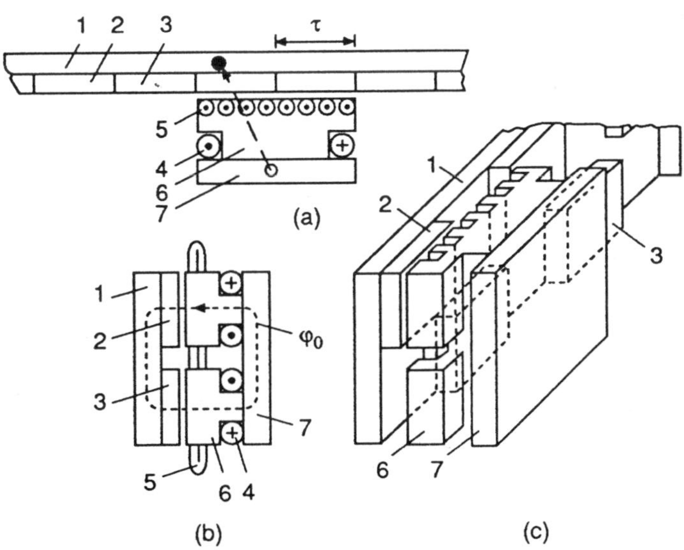

Fig. 2.28. Belikov's motor: (a) longitudinal section, (b) cross section, (c) general view, (1) ferromagnetic pole, (2) upper-row pole projections, (3) lower-row pole projections, (4) excitation winding, (5) armature winding, (6) armature magnetic circuit, and (7) yoke.

The multipole closed-end linear motor shown in Fig. 2.29 (which was developed by Lindsley[54]) provides a constant force on a bobbin regardless of the bobbin's position within the housing. The magnetic poles are evenly spaced and they are of equal size, the space between the magnetic poles is Y and the width of the poles is X. The bobbin length is $2X + 2Y$ so that it is acted on by two of the magnetic poles at any given instant in time. The stroke of the bobbin is equal to the total length for movement within the housing, which is equal to $2a(X + Y) + Y$, where a is the number of magnetic poles within the housing minus the length of the bobbin. The number of turns on the bobbin which are being acted on by two magnetic poles is not dependent on the position of the bobbin within the housing, and, therefore, a constant force is generated on the bobbin regardless of its position within the housing. The main advantage of this motor is that the efficiency of the magnetic circuit is improved, allowing a higher force to be developed for less magnetic volume.

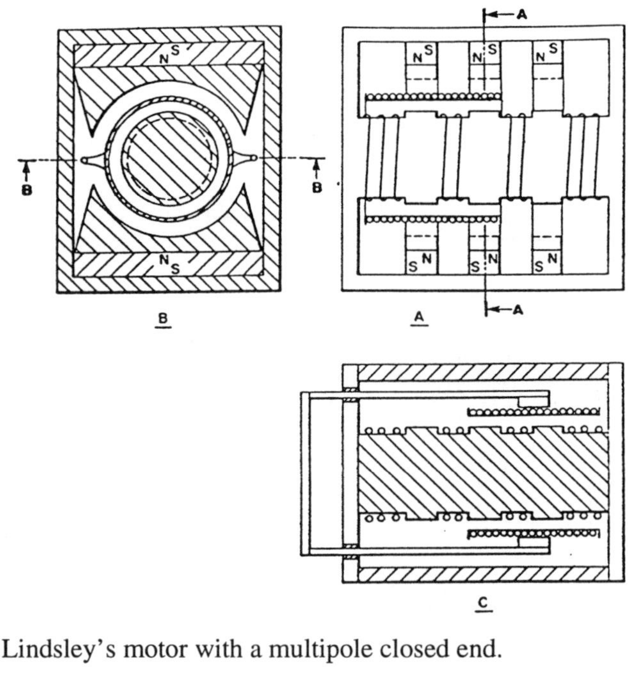

Fig. 2.29. Lindsley's motor with a multipole closed end.

A motor developed by Moczala[62] is shown in Fig. 2.30. It has a high pulling force and efficiency. Since, in this case, the stroke is short, there is no need for mechanical or electronic commutation. The motor has a similar principle of operation to the motor designed by Basak[11] in 1975 which is shown in Fig. 2.31.

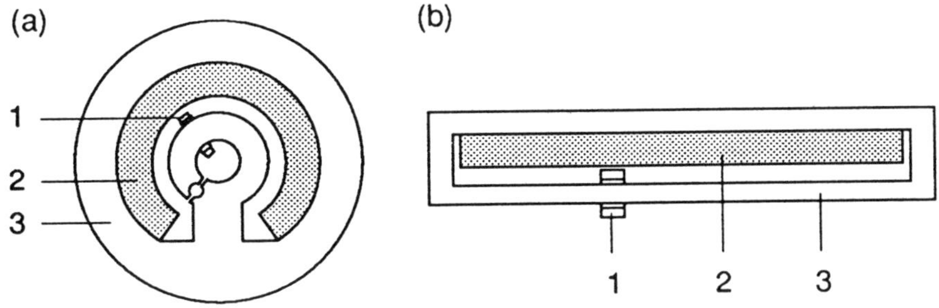

Fig. 2.30. Moczala's d.c. linear motor: (a) rotary, (b) linear (1) coil, (2) permanent magnet, and (3) soft iron yoke.

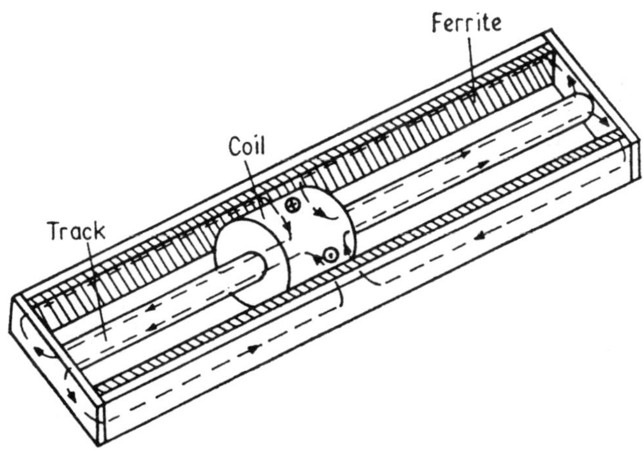

Fig. 2.31. A closed-end linear motor.

Various types of d.c. linear motor have been described in this chapter. The magnetic field in most of these recent motors is provided by permanent magnets. This is because cheap magnets, such as ferrite magnets, and high-energy magnets, such as neodymium–iron–boron magnets, are now available to suit most applications. Different types of permanent magnet, their properties, and applications are discussed in the next chapter.

3

PERMANENT MAGNETS

In the last few decades, as the energy density of permanent magnets has improved, there has been a large increase in the use of permanent-magnet materials in many nontraditional devices and also in energy-conversion systems such as motors and generators. This chapter contains a historical survey of permanent magnets, and the properties and future uses of the presently available magnets are discussed.

3.1 History

The early history of permanent magnets is rather fascinating. The only magnetic material available in ancient times was lodestone. Lodestone is a form of magnetite (Fe_3O_4) which is magnetic in its natural state. Greek philosophers of the period 100–200 BC first mentioned the attractive powers of lodestone. Good magnet steel was manufactured in China from AD 500 onwards. The first practical use of magnetism may have been in compasses, which are believed to have been introduced into Europe from China around AD 1200. In 1600, William Gilbert gave the first systematic report on permanent magnetism and on lodestones. He described three ways by which permanent magnetism could be given to iron and steel: drawing a lodestone, from the middle of a needle to the end; forging a steel specimen or drawing an iron wire pointing north and south in the earth's magnetic field; and placing a red-hot bar in the earth's magnetic field and letting it cool, or exposing an unheated bar to the earth's magnetic field for a long period. The success achieved by such methods was reasonable; In the mid-eighteenth century, Gowin Knight, for example, made a magnet which could lift 28 times its own weight. Most of these magnets were built up from wires or strips since material with a small cross section is more easily magnetized by stroking with another magnet. Special alloy steels containing tungsten were developed towards the end of the last century. Tungsten was replaced, because of its scarcity, by chromium during the First World War. Cobalt steels were discovered in Japan in 1917. Steels containing 30–40 per cent cobalt plus tungsten and chromium have a high coercivity. Alloys based on aluminium, nickel, and iron were discovered in Japan in 1931. Cobalt was added later. These metal magnets were developed for over 20 years in four categories: isotropic, anisotropic, high coercivity, and columnar.

Ceramic magnet materials were developed by the Philips organization in the 1940s. These magnets, being brittle, have a high resistivity, 10^{10} μΩ m is a typical value, compared with a value of around 0.5 μΩ m for metal magnets, and hence they have negligible eddy-current losses when they are subjected to high-frequency alternating fields. In recent years the major research interest has been in rare-earth permanent-magnet materials. Table 3.1 shows approximate discovery dates for various permanent magnets.

Table 3.1 Approximate dates of discovery of various permanent magnets

Material	Date
Lodestone	200–600 BC
Powdered iron oxide	AD 1750
Alnico (isotropic)	AD 1935
Alnico (anisotropic)	AD 1940
Barium and strontium ferrites	AD 1957
Samarium–cobalt	AD 1974
Neodymium–iron–boron	AD 1985

3.2 Properties

The relationship between the flux density (or induction), B, in a permanent magnet and the magnetizing field strength, H, may be represented by a hysteresis loop. Figure 3.1 shows the familiar hysteresis loop, or B–H characteristic, of a magnetic material. It represents a cyclic change in the magnetic flux density due to a change in magnetizing force. In Fig. 3.1, point O represents the unmagnetized state of a magnetic material and the initial magnetization curve is given by OACD. The curve AEFGHIA is a minor hysteresis loop which is obtained when the magnet is not magnetized to saturation. Point C corresponds to magnetic saturation of the material; that is, the magnet no longer contributes to any increase in the flux density. Any further increase in B is attributed to the relationship between the magnetizing field strength and the flux density of the space which is coincident with the magnet. The value of B when H=0 is given by OE and it is called the remanent

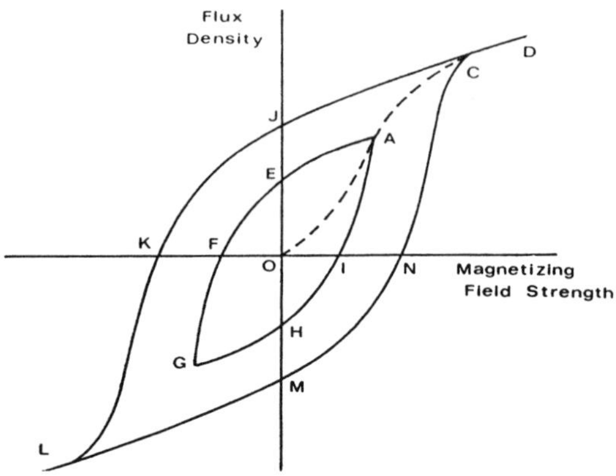

Fig. 3.1 Magnetization cycles.

magnetization, and the value of H when $B = 0$, which is given by OF, is called the coercive force.

The hysteresis loop with the largest area is generally known as the major hysteresis loop; this is shown in Fig. 3.1 by DJKLMND. If the magnet is magnetized to its saturation value B_{sat}, then the point J corresponds to the magnetization state of the material when the magnetizing force is reduced to zero. The value of the flux density at the point J is known as the remanent flux density, B_r, and it is a large proportion of the value of B_{sat}. If the magnetizing force is then increased in the negative direction the flux density reduces until it is zero at H_c, the coercive field strength. The saturation of the magnet in the reverse direction, $-B_{sat}$, is reached at point L by increasing the magnetizing force still further in the reverse direction.

By increasing the magnetizing force back to zero, the flux density reaches the value, $-B_r$. If the magnetizing force is gradually increased in the positive direction, the flux density follows the curve through zero to the value $+B_{sat}$ at point D. The loop thus obtained is the major hysteresis loop of the material.

3.2.1 *Intrinsic magnetization*

The flux density in a magnet in air is the vector sum of the intensity of magnetization (or intrinsic magnetization) M and the flux density of the space the magnet occupies. It can be expressed (in tesla) as follows:

$$B = M + \mu_0 H, \tag{3.1}$$

where μ_0 is the permeability of free space.

The variations of the flux density, B, and the intrinsic magnetization, M, with the field strength, H, are shown in Fig. 3.2. It should be observed that the intrinsic flux density, M, unlike the normal flux density, B, reaches a saturation value. The intensity of magnetization is a property of a material which is dependent upon the application of a magnetic field. For ferromagnetic materials the intensity of magnetization is large compared with the applied magnetic field, and it is nonlinearly dependent upon the applied field. The intrinsic magnetization curve is also more directly connected with the fundamental processes occurring in the magnetic material than the normal B–H curve.

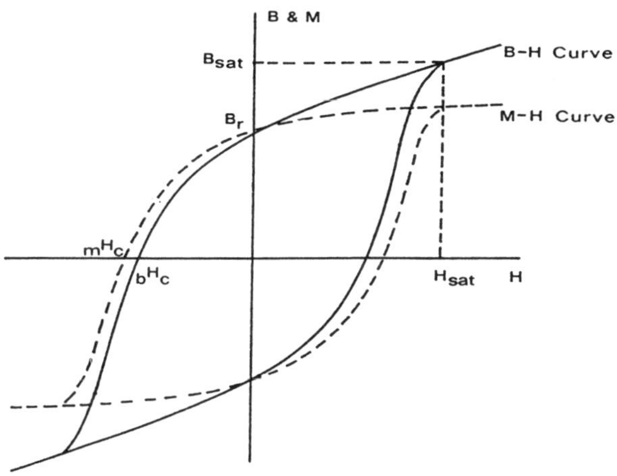

Fig. 3.2. Intrinsic magnetization.

A toroid of permanent-magnet material, magnetized to saturation, does not produce a useful external field (assuming there is no flux leakage from the toroid). The intensity of magnetization forms a continuous path in the material, and in order to produce an external field the path must be made discontinuous. A slot cut in the toroid produces an external field, H_g, in the air gap which is in the same direction as the intensity of magnetization. The air gap also creates magnetic poles and a concurrent demagnetizing field within the magnet. The internal self-demagnetizing field is directly opposite to the magnetization, M, and it is proportional to it.[35]

3.2.2 *Demagnetization characteristics*

The second and fourth quadrants of the major hysteresis loop of permanent magnets are of more interest than the other two quadrants or than the complete loop; this is because, in the absence of externally applied fields, a magnet operates, due to its own demagnetizing field, at a point in the second (or fourth) quadrant of the hysteresis loop. This part of the major hysteresis loop is called the demagnetization characteristic. The operation of a permanent magnet, and its ability to produce a flux and a magnetomotive force (m.m.f.) under various conditions of service, can be almost entirely explained in terms of this characteristic. In Fig. 3.3, the point B_r represents the magnetization state of the toroid when the magnetic field used to magnetize the toroid is reduced to zero. When an air gap is introduced, the self-demagnetizing field causes the operating point of the magnet to move to E.

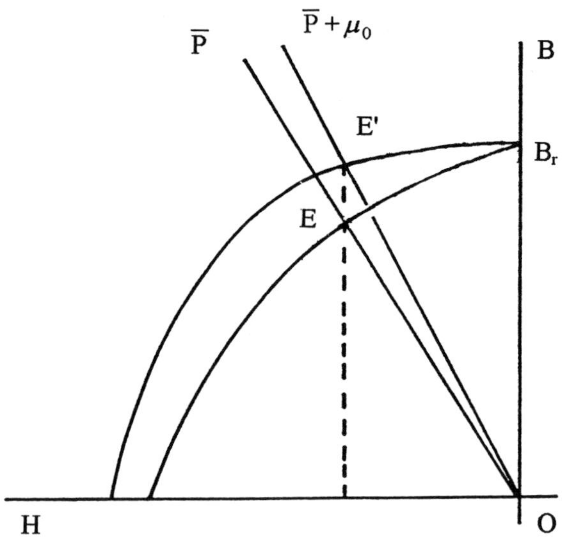

Fig. 3.3. The demagnetization characteristic.

3.2.3 *The demagnetization factor*

The demagnetization factor can be exactly determined for a completely closed circuit, using a toroid, which produces no external flux, and for a complete open circuit using an ellipsoid. For an ellipsoid, the internal demagnetizing field is constant throughout the volume, and it is proportional to the intensity of the magnetization due to its special properties. The demagnetization factor, D, can be calculated in terms of the magnet dimensions for an ellipsoidal magnet by

$$H = -\frac{M}{\mu_0} D , \tag{3.2}$$

where $0 < D < 1$

Demagnetization factors can be calculated for shapes other than ellipsoids but the values which are obtained are only approximate. The values of B, H, and M for magnet shapes other than ellipsoids vary over the magnet volume, and therefore the whole of the magnet does not work under the same conditions, that is, not at the same point on the demagnetization characteristic.

Equations (3.1) and (3.2) give

$$B = \mu_0(1-1/D)H , \tag{3.3}$$

The working point of the magnet is given by the coordinates B and H which correspond to point E in Fig. 3.3. A straight line drawn from the origin through E corresponds to the *unit-permeance* line of slope $\mu_0(1-1/D)$.

3.2.4 *Maximum energy product*

The maximum product of B and H is an important quality factor which is used to evaluate the useful properties of permanent-magnet materials. A typical demagnetization curve is drawn to the left of the ordinate in Fig. 3.4. The abscissa to the right of the ordinate is the product of B and H (taken from the demagnetization curve and plotted against B); that is, EF represents the product of DE and DG. The product at any point on the demagnetization curve indicates the external energy produced per unit volume for different points on the curve. The external energy is zero at both B_r and H_c, and it reaches a peak value at a point known as the peak energy product, $(BH)_{max}$. This point represents the maximum external energy that can be produced by a unit volume of a given material. It is the criterion which is used for comparing the values of different permanent-magnet materials. In the next section we will find that for maximum economy of permanent-magnet material the intercept of the permeance line with the B–H curve should occur where the product is a maximum, and hence for static permanent-magnet applications the volume is minimum. Figure 3.5 shows comparative sizes for various magnets as magnet performance has steadily improved. It also shows bar charts for energy products and working temperatures for different magnets. We can observe that although neodymium–iron–boron magnets have the lowest working temperatures, sintered magnets have the highest energy products of all types of permanent magnet.

3.2.5 *Magnetic circuits*

Magnetic circuits generally consist of three elements: the permanent magnets, the air paths, and the iron paths for the flux. The magnetomotive forces produced by the magnets determine the total flux in the external circuit because

$$\phi = FP = \frac{F}{\mathfrak{R}} , \tag{3.4}$$

where ϕ is the flux (in webers), F is the m.m.f., P is the magnetic permeance and $\mathfrak{R}$ is the reluctance.

The design of a magnet is mainly concerned with determining the basic dimensions and specifying a material to produce a sufficient m.m.f. to overcome the reluctance of the magnetic circuit and to produce the desired flux in the air gap.

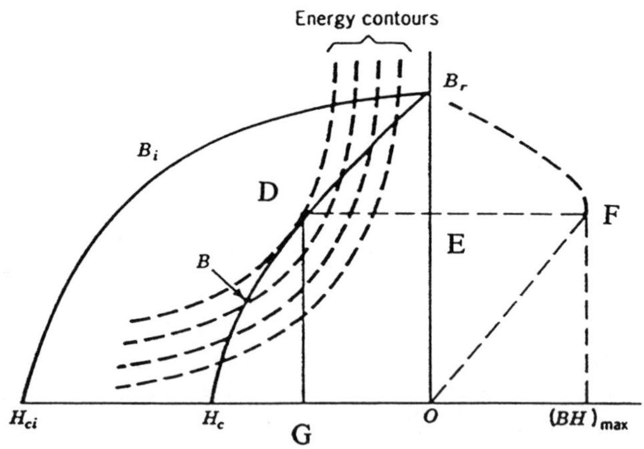

Fig. 3.4. Demagnetization curve, constant B–H contours.

If the dimensions of the gap and the flux required in it are known, the length of the magnet (in metres) may be determined with the aid of the expression

$$l_m = -\frac{qH_g l_g}{H_m} \ , \tag{3.5}$$

where the subscripts m and g denote the magnet and air gap respectively.

The reluctance factor,[23] q, is generally assumed to equal 1.1. The cross-sectional area of the magnet, A_m, may be obtained using the following expression:

$$A_m = p\frac{B_g A_g}{B_m} \ . \tag{3.6}$$

The leakage factor, p, is a minimum when the magnet is as close to the working gap as possible. Typical examples of permanent-magnet circuit configurations are shown in Fig. 3.6.

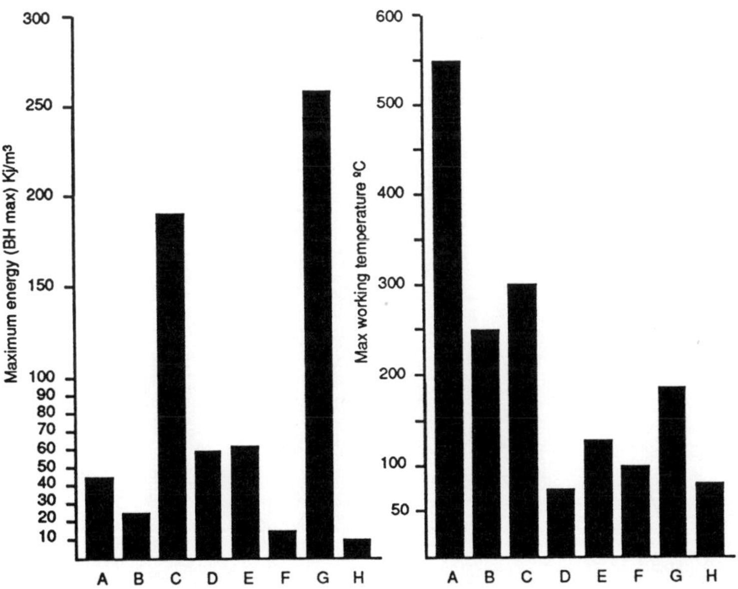

Fig. 3.5. The energy products and working temperatures of various magnets: (A) cast Alnico, (B) sintered ferrite, (C) sintered SmCo, (D) bonded SmCo, (E) bonded NdFeB, (F) bonded ferrite, (G) sintered NdFeB, and (H) flexible ferrite.

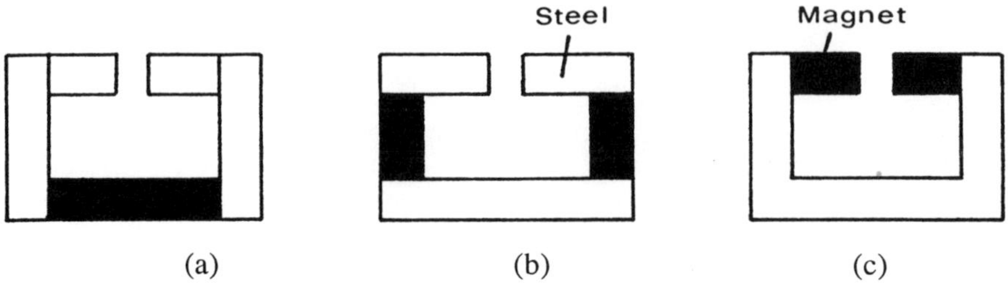

Fig. 3.6. Three different permanent-magnet configurations; configuration (c) has the lowest leakage factor.

The volume of permanent-magnet material required to produce a specified flux density in the gap can be obtained by combining eqns (3.5) and (3.6).

$$V_m = \frac{B_g^2 A_g l_g pq}{\mu_0 B_m H_m} \quad . \tag{3.7}$$

Thus, four times as much permanent-magnet material is required, if the other factors remain the same, to double the air-gap magnetic flux density. From the fundamental definitions,

$$F_g = H_g l_g \ , \tag{3.8}$$

and

$$F_m = H_m L_m \ . \tag{3.9}$$

Since the sum of the m.m.f. produced by the magnet and that required for the air gap must both be zero, by equating eqns (3.8) and (3.9), the identity

$$L_m = \frac{H_g L_g}{H_m} \tag{3.10}$$

can be obtained.

Here, L_m is the magnet length necessary for the air gap only, whereas l_m in eqn (3.5) is the actual magnet length, and it is always greater than L_m by the reluctance factor q. The reluctance factor is the ratio of the total m.m.f. produced by the magnet to the m.m.f. required for the air gap. By adding the m.m.f.s required for the air gap, the soft yoke, the m.m.f. necessary to overcome joint reluctance, and the additional m.m.f. necessary to accommodate the geometry of the air gap itself, the total m.m.f. required for the circuit can be determined. The reluctance factor can, therefore, be expressed as

$$q = 1 + \frac{F_1 + F_2 + F_3}{F_g} . \tag{3.11}$$

The ratio of the total magnetic flux in the magnet to the useful air gap flux is known as the leakage factor, p. This can be expressed as

$$p = \frac{q(1 + p_1 + p_2 + p_3)}{p_g} . \tag{3.12}$$

The location of the magnet in the circuit is also a factor in the design. Figure 3.6 shows three different configurations for the magnets and yoke. It has been found by experiment[35] that by positioning the permanent magnet close to the air gap the maximum flux density in the air gap can be obtained.

An important parameter known as the permeance coefficient, $\overline{P}$, can be developed from eqns (3.5) and (3.6):

$$\frac{B}{H} = \frac{-pA_g l_m \mu_0}{qA_m l_g} = -\overline{P} . \tag{3.13}$$

This is the equation of a straight line through the origin with a slope of –P. From eqns (3.1) and (3.13),

$$\frac{M}{H} = -(\overline{P} + \mu_0) . \tag{3.14}$$

A line of slope $-\overline{P}$ drawn from the origin intercepts the B–H curve at E and a line of slope $-(\overline{P} + \mu_0)$ intercepts the intrinsic curve at E' in Fig. 3.4.

3.2.6 *The recoil line*

Some permanent magnets in applications such as lifting work with a varying gap; hence the working point of the magnet changes under recoil conditions. This can be demonstrated by considering the case of a lifting magnet.[19] Usually, the iron armature is some distance from the magnet which attracts it. In this condition, the working point of the magnet, determined by its dimensions and by the leakage gap, is denoted by M on the demagnetization curve in Fig. 3.7. Here, MX represents the leakage flux density, and the magnetic field strength is denoted by OX. As the armature approaches the magnet the leakage flux decreases and the useful flux increases in proportion to the permeance; the reluctance of the circuit also decreases, and this increases the total flux density. Due to these changes the working point moves up the straight line MP, but it does not retrace the original demagnetization curve. The straight line MP is known as the recoil line, and its slope is usually denoted by μ_{rec}. At Q, the useful flux density is QR, and the leakage flux density drops to RS. An m.m.f. of $H_m L_m = OS \times L_m$ appears across both the useful and the leakage gaps. The area QRTY is known as the recoil product, and it represents the useful energy which is available for doing work under dynamic conditions.

When the armature touches the polepieces there is no gap, and there is no leakage flux if the iron armature is assumed to have an infinite permeability. The useful flux density is OP, and the recoil product is zero, although the force on the armature is a maximum.

The area of OPM is a triangle and from simple geometry it can be shown that the rectangular area QRTY is a maximum when

$$TY = \tfrac{1}{2} PO \qquad QY = \tfrac{1}{2} OX$$

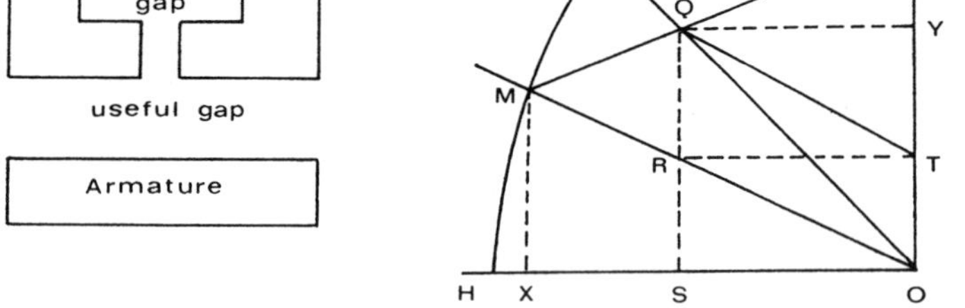

Fig.3.7. The working of a permanent magnet with full recoil.

Therefore, the maximum recoil product occurs when Q is at the midpoint of the recoil line MP.

Due to the armature reaction effect, there is an external demagnetizing field in a motor. It is a major concern in permanent-magnet motor design that the magnet will be able to withstand the self-demagnetizing field and any external demagnetizing field. In the intrinsic demagnetization curve shown in Fig. 3.8, a unit permeance coefficient line has been drawn from the origin. The intercept of the intrinsic demagnetization curve with the unit permeance line gives the magnet's working point when the armature current is zero. As the armature current increases under no-load conditions, the permeance-coefficient line shifts to the left. When the motor runs on maximum current, the armature reaction produces an external demagnetizing field, H_a. Therefore, another permeance line with slope $(B_d/H_d)+1$, originating at H_a on the field axis, can be used to obtain the extreme working point of the intrinsic flux density, B_{iw}, and the corresponding normal flux density, B_w. The demagnetizing force which reduces the remanent flux density to zero is known as coercivity H_c. The equivalent quantity for the intrinsic curve is H_{ci}.

When the demagnetizing field, H_a, is removed, the working points do not retrace the main magnetizing curves, instead they move along the respective recoil lines, to B_{im} and B_m which occur at a self-demagnetizing field H_m. The intercept of the normal recoil line with the permeance line, B_d/H_d, gives the point B_m. The difference between B_m and B_{im} is that part of the intrinsic

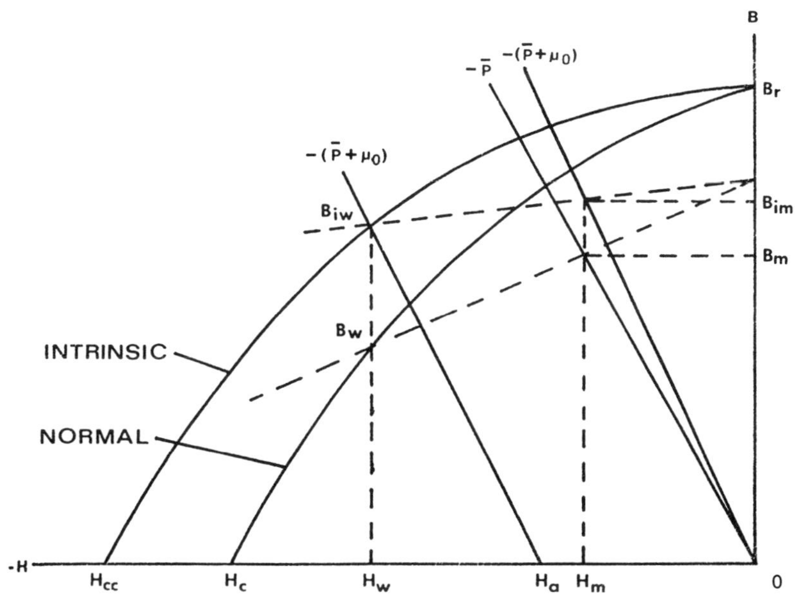

Fig. 3.8. Normal and intrinsic demagnetization curves, with the working points, in a motor.

flux density which returns from the magnet pole face through the magnet; the same proportion of B_{iw} at the extreme working point also returns through the magnet, so the effective force producing the flux density is B'_w. The difference between B'_w and B_w is the flux density produced by the m.m.f. of the armature.

3.2.7 *Stabilization*

The process of stabilizing permanent magnets implies a deliberate reduction in performance by partial demagnetization. The three main reasons for carrying out such a process are as follows:

(1) Reduction of the effects of thermal changes on operation.
(2) Protection of magnets against irreversible reduction in performance due to the presence of demagnetizing fields.
(3) Calibration of magnets for precise performance.

It has been shown[35] that magnets which have been partially demagnetized are less prone to thermal instability. To obtain stabilization, therefore, a partial reduction in performance is made when a magnet operates in fluctuating temperatures.

Again, the application of a demagnetizing field will temporarily shift the original permeance line OP_1 in Fig. 3.9 to OP_2; this moves the working point from B_3,H_3 to B_1,H_1. When the demagnetizing field is removed, the working

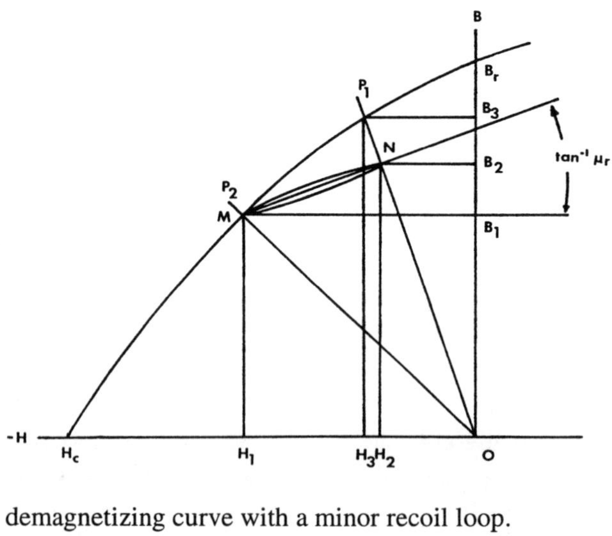

Fig. 3.9. A demagnetizing curve with a minor recoil loop.

point recoils to B_2,H_2. The effect of this treatment is that within the range H_2–H_1 , the magnet operation becomes reversible. The application of an external field with a lower value than the stabilizing field will not cause a permanent reduction in magnet performance since any variation between the limits H_2 and H_1 results in a change in the magnet flux density according to the minor loop MN.

Magnets which are calibrated for precise-performance applications, such as those used in indicating instruments, are usually given protection against the incidence of stray fields by the treatment described above. The removal of the stray field re-establishes the original calibration.

3.3 Production

The present distribution of permanent magnets in world markets is shown in Table 3.2. The total market price of magnets is over a billion pounds. Of this total, 20 per cent of the magnets are produced in Europe. Japan and the USA are the other countries which have a major share in the world market. Recently, China's share of the market has increased since its production of rare-earth permanent magnets is rapidly increasing.

Table 3.2 The present distribution of permanent magnets

Magnets	Percentage by weight
Ferrite	55%
Alnico	10%
Rare earth	35%

Currently, there is insufficient use of Alnico magnets. Motor manufacturers either use ceramic magnets for cheap motors or neodymium–iron–boron magnets for specialized compact motors. Therefore, only the production of these two types of magnet will be described in this section.

3.3.1 *Ceramic magnets*

The chemical formulation of ceramic magnets is $MO \cdot 6(Fe_2O_3)$ where M is barium, strontium, or lead. They are produced by powder metallurgy. The raw materials are first prefired and then milled to single crystals which are of the same order as the single-domain size (about 1 μm). The milled powder is then wet or dry pressed under the influence of a magnetic field (5 kOe). The pressed compacts are then sintered in air. The shrinkage is approximately 15 per cent. The critical events in making ferrite magnets are shown in Fig. 3.10. Trends in the development of the properties of this type of magnet are shown in Fig. 3.11.

3.3.2 *Neodymium–iron–boron magnets*

In 1983 Sumitomo Special Metals of Japan developed a new rare-earth iron magnet with a maximum *BH* product of 35 MGOe. It has a uniaxial structure with magnetocrystalline anisotropy, and therefore it can be easily magnetized in one direction. The anisotropy of rare-earth magnets is quite large compared with ceramic magnets. The production of these magnets can be carried out by a powder metallurgical process. Figure 3.12 shows the general steps in the production process. The main steps taken during production are described below.

Alloy preparation. The vacuum induction-melting method is used to prepare the alloy. Iron and boron are initially melted in an alumina crucible in an argon atmosphere. After degassing under a vacuum, the temperature of the molten alloy is allowed to fall to just above the temperature of the liquid, and neodymium is then added. The melt is cast into a thick-walled copper mould with a cavity of a few millimetres.

Premilling. After vacuum melting and casting, the neodymium–iron–boron alloy is in the form of chill-cast lumps. These are crushed in a nitrogen atmosphere in a high-energy hammer mill to particles which are smaller than 500 μm in diameter.

Milling. Milling is used to produce a narrow size distribution of single-crystal particles, that is, individual particles containing no grain boundaries and therefore only one preferred axis of magnetization. After milling, the powder is either dried under a vacuum, or heated in an argon atmosphere in the ball-milling method. In the case of the jet-milling method, the drying process is eliminated.

Composition control and adjustment. The magnetic properties of rare-earth permanent magnets are critically dependent on their chemical composition. Any contamination, particularly by oxidization of the alloy during processing, depletes the alloy of the rare-earth component.

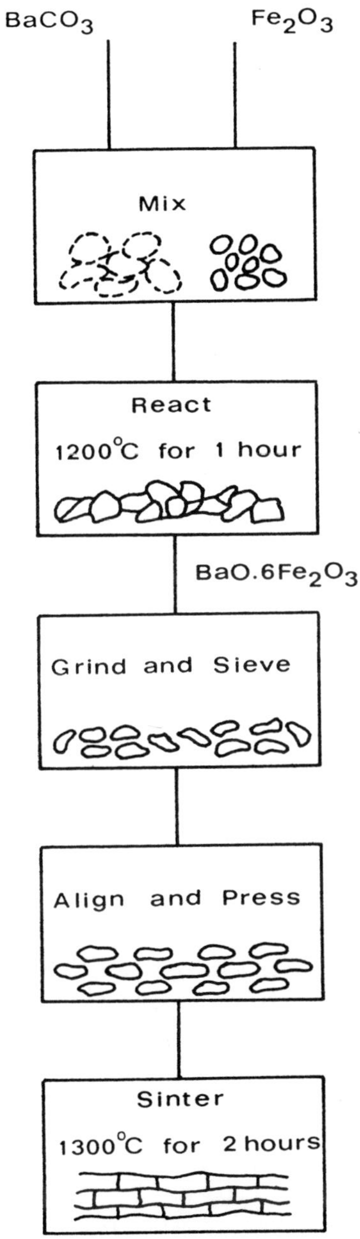

Fig. 3.10. The process of making barium ferrite.

Particle alignment and pressing. In order to obtain a compact powder with maximum magnetization, the powder particles are magnetically aligned and pressed so that the easy axes of magnetization of the particles are parallel. This powder compaction is performed by diepressing or by isostatic pressing. Isostatic pressing gives higher values of B_r and $(BH)_{max}$ than are given by diepressing.

Sintering and heat treatment. The sintering of rare-earth permanent magnets is carried out in an inert-gas atmosphere, in a reducing atmosphere, or in a vacuum. The properties of a neodymium–iron–boron magnet directly after sintering allow the magnet to be used as a permanent magnet in the as-sintered form. Figure 3.13 shows the demagnetization curve of a neodymium–iron–boron sample after sintering at 1373 K for one hour.

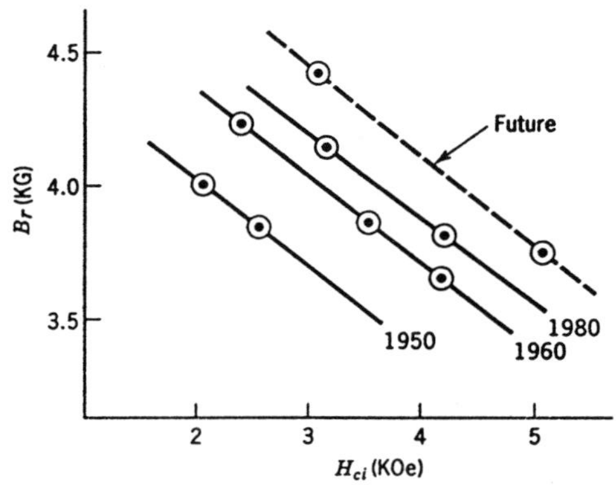

Fig. 3.11. Development trends in the properties of ferrites.

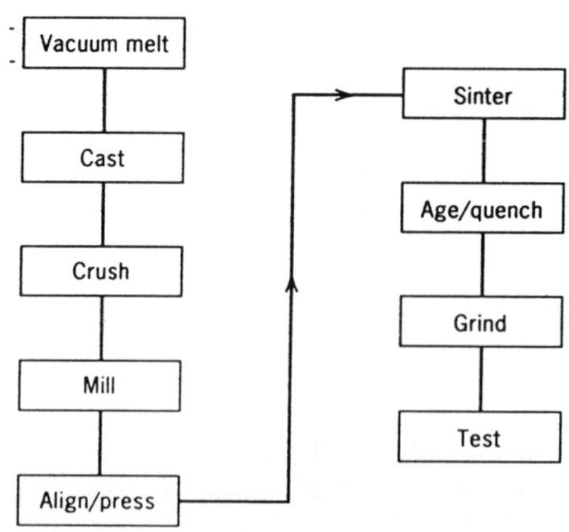

Fig. 3.12. The production of sintered neodymium–iron–boron magnets.

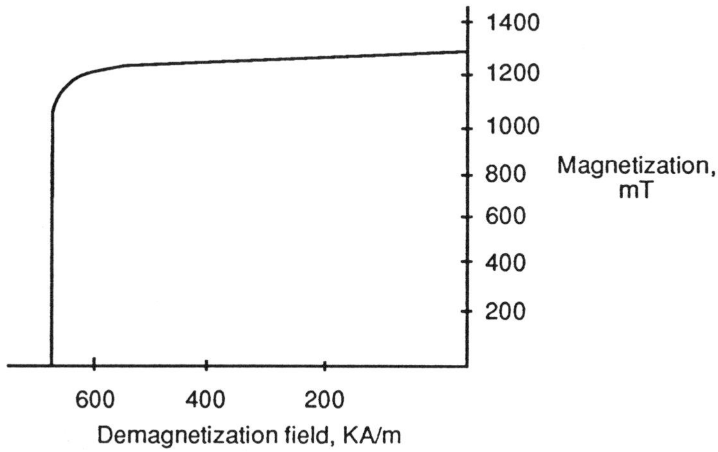

Fig. 3.13. A demagnetization curve of a neodymium–iron–boron sample after sintering at 1373 K for one hour.

Matching and magnetizing. Rare-earth permanent magnets are, in general, hard and brittle, although neodymium–iron–boron magnets are rather tougher, and they are less susceptible to breakage and chipping; machined surfaces are required either to give the necessary magnetic contact with the associated components in the final assembly or for correct size. Since most rare-earth permanent magnets combine relative recoil permeabilities which are close to unity with high coercivities, they may be magnetized prior to assembly without any flux loss.

In the next section we will discuss the application of permanent magnets, mainly in motors.

3.4 Applications

Permanent-magnet materials have a high coercive force; hence they are used to best advantage in d.c. motors, magnetic separators, and lifting devices where a high coercive force is required. Low-power and low-cost motors, which are mainly used for automobile accessories, employ these magnets extensively as segmental poles. Permanent-magnet field motors offer several advantages over conventional, wound, field rotary and linear d.c. motors. Due to the absence of a field winding, the field system occupies less space, and the motor therefore becomes more compact and lighter. The possibility of motor breakdown due to failure of the field winding naturally reduces to zero, and the motors run at much cooler temperatures. The simplified assembly requirements of motors using permanent magnets reduces their cost and increases their reliability and efficiency.

Neodymium–iron–boron magnets are the current choice of motor manufacturers worldwide. They give higher speed, power-to-weight ratio, and torque for a given size of magnet. Table 3.3 compares motors made using different types of magnet but which have the same output. However, in general, neodymium–iron–boron magnets cannot compete with ferrite magnets which are inexpensive and are made from nonstrategic raw materials. They have a high coercive force combined with a reasonable induction. Only when the volume or weight is restricted are the neodymium–iron–boron magnets the ideal choice.

Above a certain size ferrite magnets are of little use and wound-field magnets are used. Neodymium–iron–boron magnets are used in the size transition between these two other magnets. At present, there is insufficient use of Alnico magnets. It is probable that within a decade Alnico magnets will cease to be manufactured. Table 3.4 shows the cost of various magnets. The machining of the magnets into the correct shape accounts for 50 per cent of the cost. Recent increases in the price of cobalt have increased the cost of $SmCo_5$ by 15 per cent. Neodymium is known as a rare-earth material because of the difficulty in separating it from its ore. However, it is actually more abundant than lead (24 parts per million in the earth's crust compared to 16 parts of lead per million in the earth's crust).

Table 3.3 A comparison of the dimensions of a four-pole, 2500 rev min^{-1}, 2 kW brushed motor

	Ferrite FXD 460	Samarium cobalt Recoma 20	Neodymium–iron–boron Crumax 301
Magnet length (mm)	14.2	2.9	2.7
Armature length (mm)	117	93	90
Gap flux density (T)	0.2	0.5	0.56
Magnet mass (kg)	2.17	0.46	0.36
Motor mass (kg)	15.0	8.5	8.0

Table 3.4 A comparison of magnet costs (Philips Components Ltd)

RES 278 ($Nd_2Fe_{14}B$)	RES 190 ($SmCo_5$)	FXD 380($SrFe_{12}O_{19}$)
12 pence g^{-1}	21 pence g^{-1}	0.4 pence g^{-1}

The Curie temperature, and hence the working temperature, of neodymium–iron–boron magnets can be increased by adding cobalt, but this is at the expense of the magnetic properties. Again, an improvement in the coercive force can be achieved by the addition of heavy rare-earth elements such as Tb (Terbium) or Dy (Dysprosium). Neodymium–iron–boron magnets are susceptible to rusting and corrosion, and so they are usually sold with a coating of nickel.

4

THEORY OF DC LINEAR MOTORS

The principle of operation of d.c. linear motors is similar to that of conventional d.c. rotary motors. In the case of wound field motors, the flux is obtained by winding insulated copper wire on laminated cores, whereas in the case of permanent-magnet motors, permanent-magnet blocks provide the field flux.

4.1 Conventional theory

The armature and the field winding of a wound field d.c. rotary motor are shown in Fig. 4.1(a). The armature winding is wound around the x-axis, whereas the field winding is concentric with the y-axis, Fig. 4.1(b). The field winding sets up a magnetic field in the direction shown in Fig. 4.1. By the fundamental law of motor action, if a current is passed through a conductor which is perpendicular to a magnetic field, a mechanical force is exerted on the conductor in the direction perpendicular to both the armature current and the magnetic field. Therefore, a circumferential force is developed and the armature revolves around the z-axis. Similarly, in the case of linear motors, Fig. 4.2(a), a force is developed along the armature axis, that is, along the z-axis, perpendicular to both the direction of the armature current and the magnetic field. This axial force produces a linear motion in the z-direction.

When a conductor of length l carries a current I in a direction perpendicular to a magnetic field of flux density B the conductor experiences a force F perpendicular to both the direction of current flow and the magnetic field.[44] This force is given by

$$F = BIl \ . \tag{4.1}$$

Considering the case of a linear motor, the force can be calculated in the following manner, due to Griffiths and Jones.[33] This conventional theory is not rigorous, but it is included here since it was the first published analysis of d.c. linear motors. Referring to Fig. 4.3 we can rewrite eqn (4.1) as

$$F = \frac{\phi}{A_p}\left(I_a n_p\right) l_1 \tag{4.2}$$

or
$$F = \frac{\phi}{l_1 n_p / n}\left(I_a n_p\right) l_1 \,. \tag{4.3}$$

Therefore,
$$F = \phi n I_a \tag{4.4}$$

where ϕ is the radial flux (in webers) entering the armature from the pole face, A_p is the pole face area (in m^2), I_a is the armature current (in amperes), n_p is the number of conductors under a pole, l_1 is the length of a conductor under a pole (in metres), and n is the number of conductors per metre length of the armature.

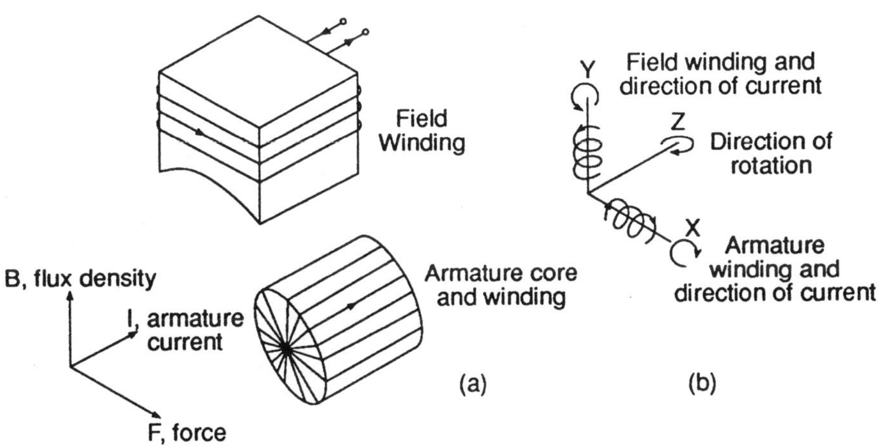

Fig. 4.1. A d.c. rotary motor.

Circular armatures. The relationship between the thrust and the armature current density, core flux density, armature core dimensions, and the armature conductor diameter can be found by using eqn (4.2).

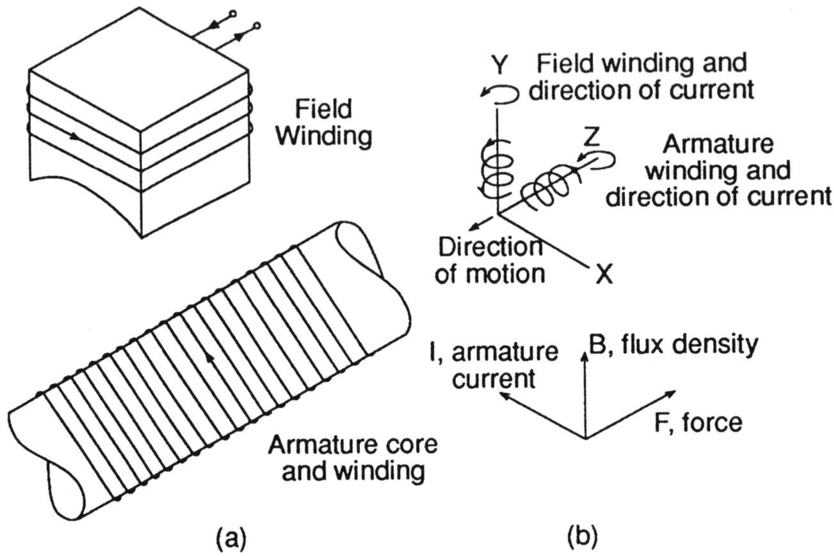

Fig. 4.2. A d.c. linear motor.

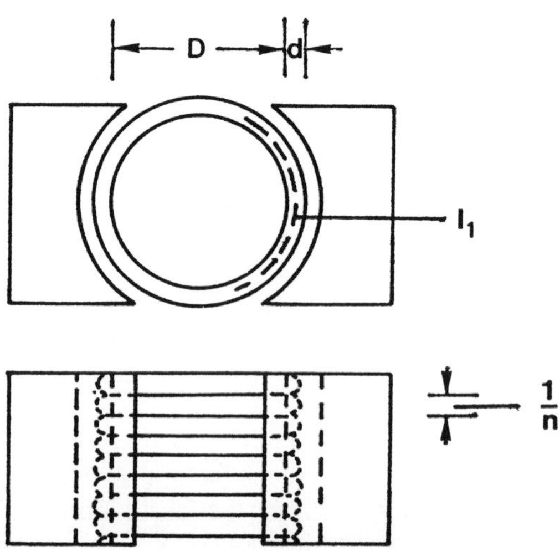

Fig. 4.3. An armature winding and field system of a linear motor.

Since

$$\phi = B_c A_c \ , \tag{4.5}$$

$$I_a = J_a a \ . \tag{4.6}$$

Again,

$$A_c = \frac{\pi}{4} D^2 \tag{4.7}$$

and

$$a = \frac{\pi \, d^2}{4} \ . \tag{4.8}$$

Therefore,

$$\phi = B_c \frac{\pi}{4} D^2 \tag{4.9}$$

and

$$I_a = J_a \frac{\pi \, d^2}{4} \ , \tag{4.10}$$

where B_c is the armature core mean flux density (in tesla), A_c is the armature core cross-sectional area (in m^2), J_a is the current density in armature conductors (in Am^{-2}), a is the conductor cross-sectional area (in m^2), D is the diameter of the armature core (in metres), and d is the diameter of the conductor (in metres).

Since n is the number of conductors per metre length,

$$n = \frac{1}{d} \ . \tag{4.11}$$

Substituting eqns (4.9), (4.10) and (4.11) in eqn (4.4) and introducing k, Jones[43] found the thrust per pole to be

$$F = k\left(B_c \frac{\pi}{4} D^2 \right)\left(J_a \frac{\pi \, d^2}{4} \frac{1}{d} \right) = \frac{k B_c D^2 J_a \pi^2 d}{4 \times 4} \ , \tag{4.12}$$

where k is the ratio of the conductor (copper) diameter to the overall wire diameter, which is known as the winding factor. As the conductor diameter increases, k approaches unity, since the insulation thickness becomes negligible.

The total thrust for a two-pole motor with an armature of circular cross section becomes:

$$F_{\mathrm{t}} = 2\left(\phi\ nI_{\mathrm{a}}\right) = \frac{B_{\mathrm{c}} D^2 k J_{\mathrm{a}} \pi^{2} d}{8}\ . \tag{4.13}$$

Rectangular armatures. For a linear motor with an armature core with a rectangular cross section:

$$\phi = B_{\mathrm{c}} D_1 D_2\ , \tag{4.14}$$

where D_1 and D_2 are the breadth and width of the armature core.

Again,
$$n = \frac{1}{d}\ , \tag{4.15}$$

$$I_{\mathrm{a}} = J_{\mathrm{a}} \frac{\pi\ d^2}{4}\ ; \tag{4.16}$$

therefore,
$$F = \frac{B_{\mathrm{c}} D_1 D_2 J_{\mathrm{a}} \pi\ d}{4} \tag{4.17}$$

and the total thrust for two poles is

$$F_{\mathrm{t}} = \frac{B_{\mathrm{c}} D_1 D_2 k J_{\mathrm{a}} \pi\ d}{2}\ . \tag{4.18}$$

If the armature core has a square cross section then $D_1 = D_2 = D$, and

$$F_{\mathrm{t}} = \frac{B_{\mathrm{c}} D^2 k J_{\mathrm{a}} \pi\ d}{2}\ . \tag{4.19}$$

Square armatures with two back cores. In the case of motors with square armature cores and two additional stationary back cores of the same length as the armature core (as in Fig. 2.24), the thrust can be shown to be given by

$$F = I_{\mathrm{a}} \frac{d\psi}{dx}\ , \tag{4.20}$$

where I_a is the armature current, ψ is the flux linkage produced by integrating the flux from the permanent magnets for each winding over the length of the coil, and x is the coordinate defining the position of the permanent magnets.

If we now calculate the flux linkage, ψ, and then differentiate it with respect to x, then

$$F = 2I_a(B_E + B_G)\frac{l_m N}{l_t} b \,, \tag{4.21}$$

where B_E is the flux density of the effective field, N is the number of turns of the armature winding, b is the dimension of the cores perpendicular to the plane of the diagram, B_G is the return air gap flux density, l_m is the magnet length, and l_t is the length of the cores (or the travel length).

If $l_t > l_m$ we can ignore B_G and eqn (4.16) reduces to

$$F = 2B_E I_a N \frac{l_m}{l_t} b \,, \tag{4.22}$$

$$= 2B_E b l_m \frac{N}{l_t} I_a \,. \tag{4.23}$$

Double rectangular armatures. The force developed by a d.c. linear motor built with two parallel static rectangular armatures and a slider with permanent magnets, Fig. 2.22, can be deduced from eqn (4.4); that is, from

$$F = \phi \, nI_a \,. \tag{4.4}$$

Therefore,

$$F = B_g l_1 w_m \frac{N}{l_t} I_a$$

$$= \frac{N w_m}{l_t} I_a l_1 B_g \,. \tag{4.24}$$

where N is the total number of turns of armature winding, w_m is the width of the magnets, l_t is the travel length, I_a is the armature current, l_1 is the active conductor length for each side of the coil, B_g is the magnetic flux density in the air gap.

This force is due to a single magnet located at one side of one armature. Since only one section of the armature winding is energized at any particular time and since this section is surrounded by four magnets with similar dimensions, the total force is

$$F_t = 4\frac{Nw_m}{l_t} I_a l_1 B_g \,. \tag{4.25}$$

This expression is ideal because it assumes that the air-gap flux density is uniform and that the magnetic fluxes provided by all the magnets are the same despite the topology of the magnetic circuit.

The expressions for the slider velocity, v, and the electromagnetic power, P_e, are:

$$v = \frac{V_t - Ir_a}{4(Nw_m / l_t)B_g l_1} \,, \tag{4.26}$$

and

$$P_e = 4(Nw_m / l_t)B_g l_1 v I_a \,. \tag{4.27}$$

In eqn (4.26), V_t is the terminal voltage supplied to the armature winding, and r_a is the ohmic resistance of the winding.

Readers should note that at any particular moment only one section of the armature winding is energized, and the length of this section is roughly equal to the magnet length; otherwise forces will be developed in opposite directions, and the motor will stall as a result. The magnet blocks with their south poles facing the armature do not develop any force due to the absence of any current in the conductors in these sections; see Fig. 4.4.

The dashed lines in Fig. 4.4 show the flux pattern produced by the permanent-magnet blocks in a two-pole d.c. linear motor. In this arrangement there are two separate magnetic circuits, one on the right and the other on the left of the armature core; each circuit consists of two magnetic poles, the yokes, the air gaps, the armature winding, and the core. In the first magnetic circuit, the flux from the north pole face crosses the air gap A, cuts the armature conductors, passes along the armature core, crosses the air gap B and reaches the south pole face of the other magnet block; the flux from the north pole face of the second magnet block passes along the steel yoke to the south pole face of the first magnet block.

The second magnetic circuit is similiar to the first except that the flux lines perpendicular to the armature axis are in the opposite direction. When the current flows in the armature winding in the direction shown in Fig. 4.5, each of the two magnetic circuits produces a force in the same direction, since the current and the magnetic field are in opposite directions in the two circuits.

In general, the armature is mechanically fixed to the base and therefore it forms the stationary part of the motor, whereas the permanent-magnet blocks are mounted inside the moving unit of the motor. Hence there is a movement of the field system relative to the armature.

The forces produced by the two magnetic circuits are in the same direction and they are equal in magnitude, thus a balanced linear motion of the carriage is produced. Any change in the direction of the current or in the polarity of the magnet blocks reverses the direction of the developed force.

The cross section of the armature core may have different shapes (circular, rectangular, square). The most convenient way of winding is, however, to wind a coil over a bar with a circular cross section.

Although neodymium–iron–boron permanent magnets are machinable, an arc-shaped magnet is difficult to make. In order to maintain a uniform air gap between the magnet poles and the armature winding, either of two methods can be employed. One method[46] uses a mild-steel bar of circular cross section as the armature core with the pole shoes shown in Fig. 4.6. The pole shoes are made from mild steel and they are attached to permanent magnets to collect the flux and then to transfer it to the air gap, and at the same time they allow the flux produced by the armature current to pass through them. The alternative method is to use a mild-steel bar with a square cross section as the armature core. In both cases the permanent magnets can have flat faces.

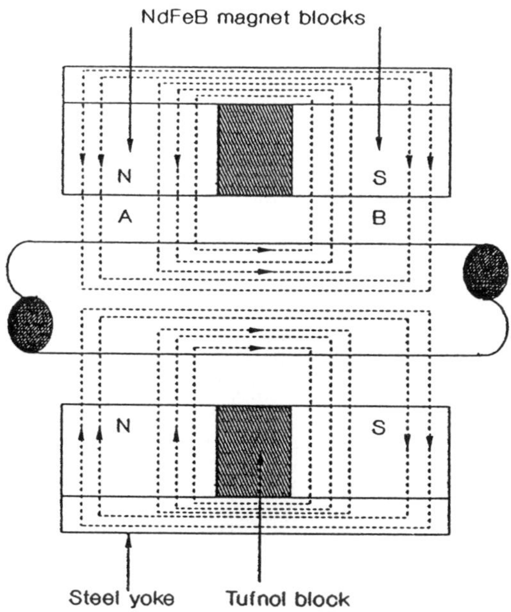

Fig. 4.4. The flux pattern due to the permanent magnets in a two-pole d.c. linear motor (not to scale).

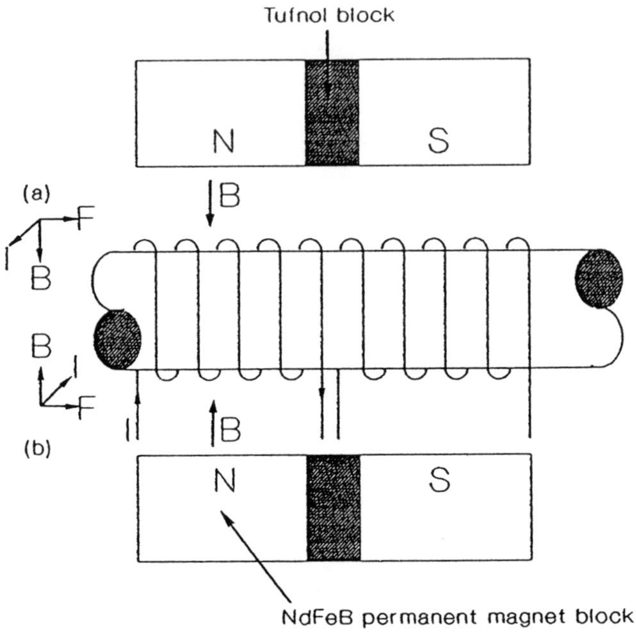

Fig. 4.5. The force development in a two-pole d.c. linear motor (not to scale).

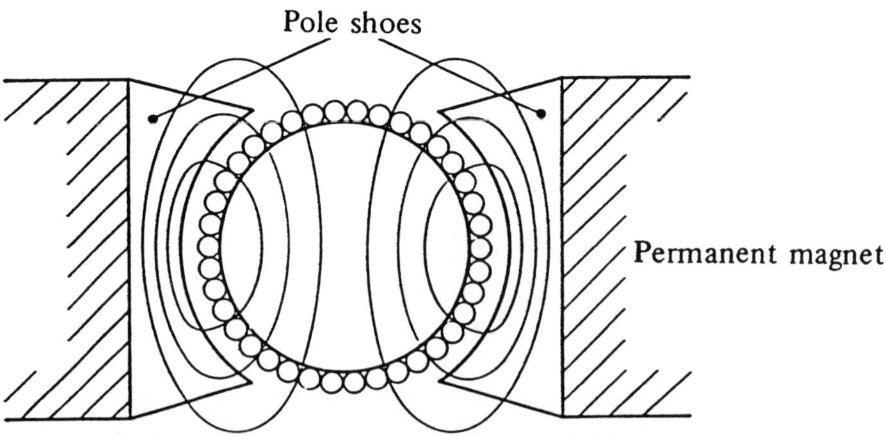

Fig. 4.6. An armature flux passing through two pole shoes.

When a conductor of length l moves with a velocity v in a magnetic field of flux density B, a voltage is induced in the conductor and the directions of B, l , and v are mutually perpendicular. This voltage is given by

$$E = Blv \ . \tag{4.28}$$

The total voltage E_t induced in the coil with a relative velocity v under a magnetic pole is given by

$$E_t = n_p E = n_p Blv \tag{4.29}$$

or

$$E_t = n_p \frac{\phi}{n_p l_1 / n} l_1 v \ . \tag{4.30}$$

Therefore,

$$E_t = \phi \ nv \ . \tag{4.31}$$

The usual expression for the efficiency of the system is

$$\eta = \frac{output \ \ power}{input \ \ power} \times 100\% \tag{4.32}$$

The input power can readily be determined from the input voltage and the current, whereas the ouput power is calculated from a knowledge of the output force and the velocity of the system.

4.2 Generalized machine theory

4.2.1 *Wound field direct-current linear motors*

The voltage equation. In general, electromechanical transducers use magnetic fields which are established by current-carrying coils. The electrical circuits then contain self-inductances and mutual inductances. If a voltage e is applied to a coil (see Fig. 4.7) then a current i flows, and a magnetic flux ϕ is produced. They are related by,

$$e = Ri + N\frac{d\phi}{dt} = Ri + \frac{d\psi}{dt} \ , \tag{4.33}$$

where R is the resistance of the coil, N is the number of turns on the coil, $\psi = N\phi$ is the total flux linkage.

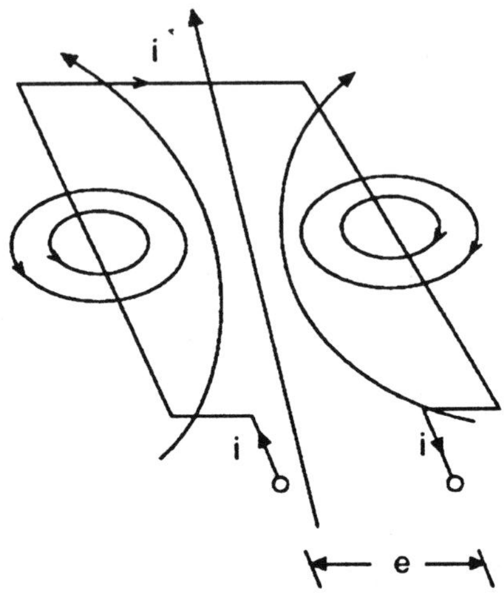

Fig. 4.7. A magnetic flux produced by current-carrying conductors.

The voltage equations can be written in terms of the terminal voltages using eqn (4.33), as follows:[70]

$$e_{\mathrm{f}}(t) = R_{\mathrm{f}} i_{\mathrm{f}}(t) + D\psi_{\mathrm{f}}(t) , \tag{4.34}$$

$$e_{\mathrm{a}}(t) = R_{\mathrm{a}} i_{\mathrm{a}}(t) + D\psi_{\mathrm{a}}(t) , \tag{4.35}$$

where

$$\psi_{\mathrm{f}}(t) = L_{\mathrm{f}} i_{\mathrm{f}}(t) + M_{\mathrm{fa}} i_{\mathrm{a}}(t) , \tag{4.36}$$

$$\psi_{\mathrm{a}}(t) = L_{\mathrm{a}} i_{\mathrm{a}}(t) + M_{\mathrm{af}} i_{\mathrm{f}}(t) , \tag{4.37}$$

and L_{f} and L_{a} are self-inductances, M_{fa} and M_{af} are mutual inductances, and the subscripts f and a denote field and armature. Assuming that saturation effects are negligible, the self-inductances L_{f} and L_{a} can be considered to be constants. The mutual inductances vary with the displacement *x*, and the currents vary with the time *t*. Therefore, the following expressions for the field and armature voltages can be derived from eqns (4.35) and (4.37):

$$e_{\mathrm{f}}(t) = R_{\mathrm{f}} i_{\mathrm{f}}(t) + L_{\mathrm{f}} \frac{di_{\mathrm{f}}(t)}{dt} + M_{\mathrm{fa}} \frac{di_{\mathrm{a}}(t)}{dt} + \left(i_{\mathrm{a}}(t) \frac{dM_{\mathrm{fa}}}{dx} \right) \frac{dx}{dt} , \tag{4.38}$$

$$e_{\mathrm{a}}(t) = R_{\mathrm{a}} i_{\mathrm{a}}(t) + L_{\mathrm{a}} \frac{di_{\mathrm{a}}(t)}{dt} + M_{\mathrm{af}} \frac{di_{f}(t)}{dt} + \left(i_{\mathrm{f}}(t) \frac{dM_{\mathrm{af}}}{dx} \right) \frac{dx}{dt} . \tag{4.39}$$

Referring to Fig. 4.2, the field winding is located on the y-axis, and the stationary armature, with a distributed winding, is represented by an equivalent concentrated winding on the z-axis. There is no coupling between the magnetic field due to $i_{\mathrm{f}}(t)$ and the magnetic field due to $i_{\mathrm{a}}(t)$; this is because their axes are in quadrature whether the field unit is in motion or is stationary. Consequently, the theoretical mutual inductances between the two windings are equal to zero,[60] and they are found experimentally to be negligible. Therefore, assuming that the field unit is stationary, any voltage that occurs will be due to resistances and self-inductances. Hence, eqn (4.39) can be written in the much simpler form,

$$\begin{bmatrix} e_{\mathrm{f}}(t) \\ e_{\mathrm{a}}(t) \end{bmatrix}_{\mathrm{STATIONARY}} = \begin{bmatrix} R_{\mathrm{f}} + L_{\mathrm{f}} D & 0 \\ 0 & R_{\mathrm{a}} + L_{\mathrm{a}} D \end{bmatrix} \begin{bmatrix} i_{\mathrm{f}}(t) \\ i_{\mathrm{a}}(t) \end{bmatrix} . \tag{4.40}$$

As soon as the field unit starts moving, a motional voltage is developed in the armature winding by $i_{\mathrm{f}}(t)$. This voltage, also known as the back electromotive force (e.m.f). is proportional to the speed of the field system $v(t)$, and it can be derived as follows:

$$e_{\mathrm{b}}(t) = \frac{d\psi_{\mathrm{a}}(t)}{dt} = i_{\mathrm{f}}(t) \frac{dM_{\mathrm{af}}}{dt}$$

$$= i_{\mathrm{f}}(t) \frac{dM_{\mathrm{af}}}{dx} \frac{dx}{dt} . \tag{4.41}$$

Therefore, $$e_{\mathrm{b}}(t) = i_{\mathrm{f}}(t) L_{\mathrm{m}} v(t) , \tag{4.42}$$

where $$L_{\mathrm{m}} = \frac{dM_{\mathrm{af}}}{dx} . \tag{4.43}$$

L_m has units of Hm^{-1},and, since it appears in the derivation of the motional e.m.f., it is called the motional inductance coefficient.

The field winding and the magnetic field due to the armature current are stationary with respect to each other during the motion; therefore, no motional e.m.f. develops in the y-axis winding (see Fig. 4.2) due to $i_a(t)$. Therefore, the final form of the voltage equations is

$$\begin{bmatrix} e_f(t) \\ e_a(t) \end{bmatrix} = \begin{bmatrix} R_f + L_f D & 0 \\ L_m v(t) & R_a + L_a D \end{bmatrix} \begin{bmatrix} i_f(t) \\ i_a(t) \end{bmatrix}. \tag{4.44}$$

This matrix equation allows the equivalent circuits of a d.c. linear motor at standstill and under dynamic conditions to be drawn (see Fig. 4.8).

The force equation. When the armature and field windings are energized, a force is developed which causes the field unit to move. This force can be deduced directly from the power input, P_{in} , which is given in matrix form by

$$P_{in} = e_f(t) i_f(t) + e_a(t) i(t) \tag{4.45}$$

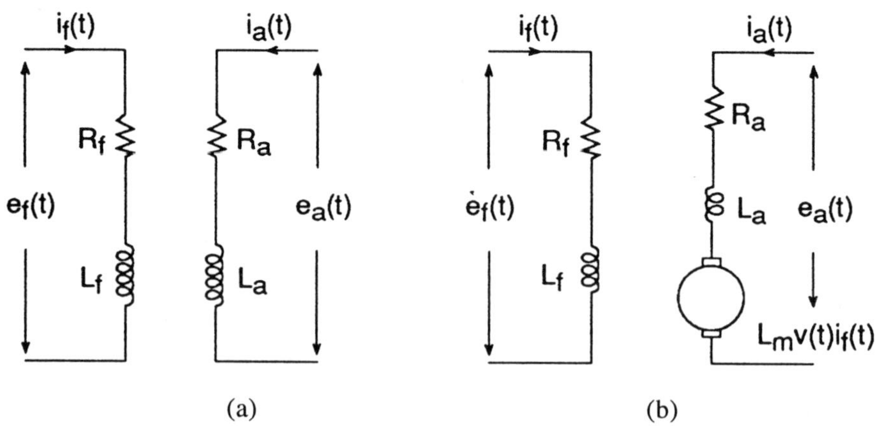

Fig. 4.8. The equivalent circuit of a wound field d.c. linear motor: (a) standstill, and (b) dynamic.

or by

$$P_{in} = [I]^T[z_0][I] \,, \tag{4.46}$$

where $[I]^T$ is the transpose of $[I]$, $[I]$ is the matrix form of the field and armature currents, $[z_0]$ is the impedance matrix in eqn (4.44). Thus

$$P_{in} = \begin{bmatrix} i_f(t) & i_a(t) \end{bmatrix} \begin{bmatrix} R_f + L_f D & 0 \\ L_m v(t) & R_a + L_a D \end{bmatrix} \begin{bmatrix} i_f(t) \\ i_a(t) \end{bmatrix} \tag{4.47}$$

or

$$P_{in} = [I]^T[R][I] + [I]^T[LD][I] + [I]^T[L_m v(t)][I] \,. \tag{4.48}$$

The first term of eqn (4.48) represents the resistance losses, and the second term is the pulsating power, which is zero for steady-state d.c. conditions. The last term represents the output power, P, which produces motion. Now the developed force is given by

$$f_d(t) = P/v(t) = [I]^T[L_m][I] \tag{4.49}$$

or by

$$f_d(t) = \begin{bmatrix} i_f(t) & i_a(t) \end{bmatrix} \begin{bmatrix} 0 & 0 \\ L_m & 0 \end{bmatrix} \begin{bmatrix} i_f(t) \\ i_a(t) \end{bmatrix} \tag{4.50}$$

or by

$$f_d(t) = i_f(t) L_m i_a(t) \,. \tag{4.51}$$

The dynamic equation. The force absorbed by the moving unit includes an effect due to mass, M, a frictional force with a frictional coefficient, F, and a force, f, corresponding to a mechanical load which can be driven by the motor. By balancing the developed force, $f_d(t)$, against the absorbed force, the dynamic equation can be written as

$$f_d(t) = MD^2x + FDx + f \,. \tag{4.52}$$

This equation can be rewritten as

$$f_d(t) = MDv(t) + Fv(t) + f \,. \tag{4.53}$$

4.2.2 *Permanent-magnet field direct-current linear motors*

The magnetic circuit of the field unit of a permanent-magnet d.c. linear motor can be represented by the equation

$$F(t) = \phi\,(t)\Re \tag{4.54}$$

where $F(t)$ is the magnetomotive force, $\phi\,(t)$ is the product of the flux density at the working point and the pole area, and $\Re$ is the reluctance of the circuit.

The armature voltage equation is given by

$$e_{\mathrm{a}}(t) = \phi\,(t)nv(t) + (R_{\mathrm{a}} + L_{\mathrm{a}}D)i_{\mathrm{a}}(t) \tag{4.55}$$

where $\phi\,(t)nv(t)$ is the back e.m.f.; this is identical to the term $i_{\mathrm{f}}(t)L_{\mathrm{m}}v(t)$ in eqn (4.42).

Equations (4.54) and (4.55) can be written in the matrix form[12]

$$\begin{bmatrix} F(t) \\ e_{\mathrm{a}}(t) \end{bmatrix} = \begin{bmatrix} \Re & 0 \\ nv(t) & R_{\mathrm{a}} + L_{\mathrm{a}}D \end{bmatrix} \begin{bmatrix} \phi\,(t) \\ i_{\mathrm{a}}(t) \end{bmatrix} . \tag{4.56}$$

The electrical power which is converted to produce motion is

$$P = e_{\mathrm{b}}(t)i_{\mathrm{a}}(t) \tag{4.57}$$

or

$$P = \phi\,(t)nv(t)i_{\mathrm{a}}(t)\,. \tag{4.58}$$

Therefore, the developed force is given by

$$f_{\mathrm{d}}(t) = \frac{P}{v(t)} = \phi\,(t)ni_{\mathrm{a}}(t)\,. \tag{4.59}$$

By balancing the developed force, $f_{\mathrm{d}}(t)$, with the force absorbed by the motor due to the mass M, the friction F, and the mechanical load f, the dynamic equation can be written as

$$f_{\mathrm{d}}(t) = (MD + F)v(t) + f\,. \tag{4.60}$$

Steady state. For steady-state conditions, the following relationships can be obtained from eqns (4.55), (4.59), and (4.60):

$$e_{\mathrm{a}_0} = \phi_0 n v_0 + R_{\mathrm{a}} i_{\mathrm{a}_0} , \tag{4.61}$$

$$f_{\mathrm{d}_0} = \phi_0 n i_{\mathrm{a}_0} , \tag{4.62}$$

$$f_{\mathrm{d}_0} = F v_0 + f \ . \tag{4.63}$$

Eliminating i_{a_0} and f_{d_0} from these equations, and assuming $\phi_0 n = K$, then the steady-state velocity of a permanent-magnet d.c. linear motor is given by

$$v_0 = \frac{K\ I_{\mathrm{a}_0} - f}{F_{\mathrm{eff}}} , \tag{4.64}$$

where $F_{\mathrm{eff}} = F + K^2 / R_{\mathrm{a}}$ is the effective frictional coefficient.

Transient state. The analysis of the transient response of motors under constant field sources is more complex; therefore the effect of armature inductance will be ignored to simplify the transient analysis equations. The relations are

$$e_{\mathrm{a}}(t) = Kv(t) + R_{\mathrm{a}} i_{\mathrm{a}}(t) , \tag{4.65}$$

$$f_{\mathrm{d}}(t) = Ki_{\mathrm{a}}(t) , \tag{4.66}$$

$$f_{\mathrm{d}}(t) = (MD + F)v(t) + f \ . \tag{4.67}$$

Eliminating $i_{\mathrm{a}}(t)$ and $f_{\mathrm{d}}(t)$ from these equations gives

$$v(t)\left[MD + \left(F + \frac{K^2}{R_{\mathrm{a}}}\right)\right] = K\ I_{\mathrm{a}}(t) - f \tag{4.68}$$

or

$$\left(1 + \frac{M}{F_{\mathrm{eff}}} D\right) v(t) = \frac{K\ I_{\mathrm{a}}(t) - f}{F_{\mathrm{eff}}} \tag{4.69}$$

or

$$\left(1+\tau_{\mathrm{m}}D\right)v(t)=\frac{K\,I_{\mathrm{a}}(t)-f}{F_{\mathrm{eff}}}\,, \tag{4.70}$$

where τ_{m} is the effective time constant of the motor. The transient solution of this transfer function can be obtained by putting

$$\left(1+\tau_{\mathrm{m}}D\right)v(t)=0$$

or

$$v_{\mathrm{t}}=A\,\mathrm{e}^{-t/\tau_{\mathrm{m}}}\,, \tag{4.71}$$

where A is a constant. According to eqn (4.64), the steady-state value is given by

$$v_0=\frac{K\,I_{\mathrm{a}_0}-f}{F_{\mathrm{eff}}}\,. \tag{4.72}$$

Therefore, the transient response is given by

$$v(t)=v_{\mathrm{t}}+v_0=A\,\mathrm{e}^{-t/\tau_{\mathrm{m}}}+\frac{K\,I_{\mathrm{a}_0}-f}{F_{\mathrm{eff}}}\,. \tag{4.73}$$

At $t=0$, $v(t)=0$, so eqn (4.73) reduces to

$$v(t)=\frac{K\,I_{\mathrm{a}_0}-f}{F_{\mathrm{eff}}}\left(1-\mathrm{e}^{-t/\tau_{\mathrm{m}}}\right). \tag{4.74}$$

Figure 4.9 shows a mechanical load of mass m_l attached to a carriage of mass m. For this configuration, the values of f and F_{eff} in eqn (4.74) correspond to $m_l g$ and $\left(m+m_l\right)/\tau_{\mathrm{m}}$, respectively, where g is the gravitational constant and τ_{m} is the motor's time constant.

When the armature inductance is included in the analysis of the speed, the response becomes more complicated and Laplace operations have to be used. Taking Laplace transforms of both sides of eqn (4.55), and assuming zero initial conditions, then

$$E_{\mathrm{a}}(s)=\left(R_{\mathrm{a}}+L_{\mathrm{a}}s\right)I_{\mathrm{a}}(s)+KV(s)\,, \tag{4.75}$$

$$F_{\mathrm{d}}(s)=K\,I_{\mathrm{a}}(s)\,, \tag{4.76}$$

$$F_d(s) = (Ms + F)V(s) \ . \tag{4.77}$$

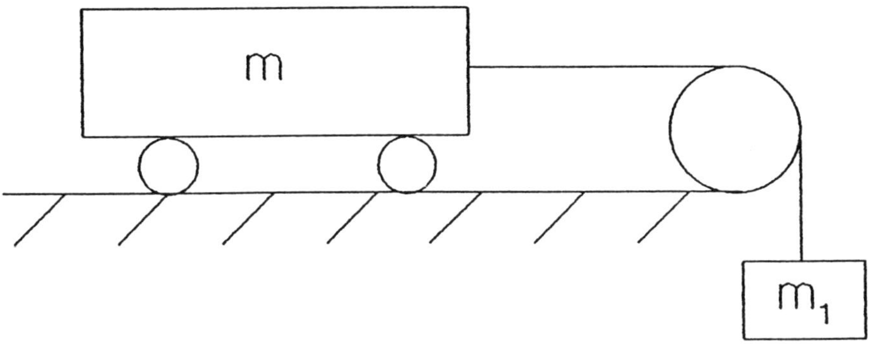

Fig. 4.9. The mechanical load attached to the motor.

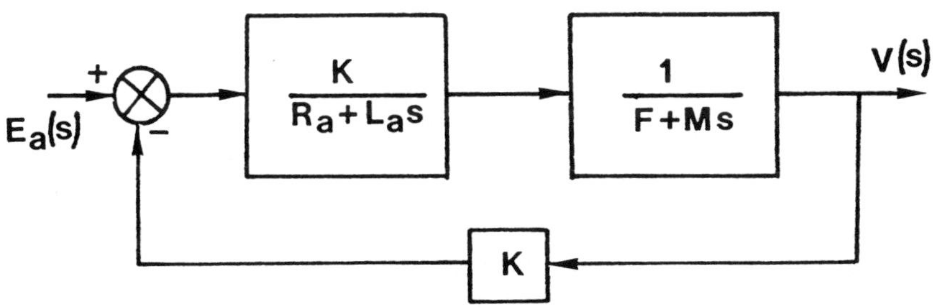

Fig. 4.10. A block diagram for an armature-controlled d.c. linear motor.

Combining these three equations, and eliminating $I_a(s)$ and $F_d(s)$, then

$$\frac{V(s)}{E_a(s)} = \frac{K}{(Ms+F)(R_a+L_a s)+K^2} ,$$

$$= \frac{K/R_a F}{(1+s\tau_m)(1+s\tau_e)+K^2/R_a F} , \tag{4.78}$$

where the motor's time constant $\tau_m = M/F$ and the electrical time constant $\tau_e = L_a/R_a$.

By employing eqn (4.78), a block diagram of the transient response of a permanent-magnet d.c. linear motor can be drawn, as is shown in Fig. 4.10.

4.3 Representation as magnetic circuits

The purpose of this section is to discuss the operation of magnets in terms of the magnet characteristics and in the context of the published work of Hanrahan and Toffolo,[37,38] and Desmond.[26]

The magnets of a permanent-magnet motor, as in other electromagnetic devices, operate in the second quadrant of the *B–H* characteristic. A typical demagnetization curve related to the *B–H* curve by the magnet dimensions (magnet length and pole area) is shown in Fig. 4.11; ϕ_r is the residual flux and F_c is the coercive m.m.f.

The flux paths which exist in permanent-magnet field d.c. linear motors are shown in Fig. 4.12. The flux, ϕ, emanating from the ferrite permanent magnets divides into a useful or air-gap flux, ϕ_u, and the leakage flux, ϕ_l. The useful flux links the armature winding, whereas the leakage paths link two adjacent but opposite poles (see Fig. 4.12). The machine may be represented by a magnetic circuit, as is shown in Fig. 4.13, where the useful path is in parallel with the leakage path.

The magnets are stabilized to ensure consistent performance by demagnetizing to a point, say (F_d , ϕ_d) as shown in Fig. 4.11. This point determines the slope of the recoil loop, which is assumed to be a straight line. The working point of the magnet is on this line, and the open-circuit m.m.f. F_o is determined by the intersection of the recoil line with the flux-axis.

From Fig. 4.13, by equating the magnetic-potential drops to the m.m.f., when the armature current is zero, we can write

$$F = -Hl = H_l l_l = H_u l_u , \tag{4.79}$$

$$\phi = \phi_u + \phi_l . \tag{4.80}$$

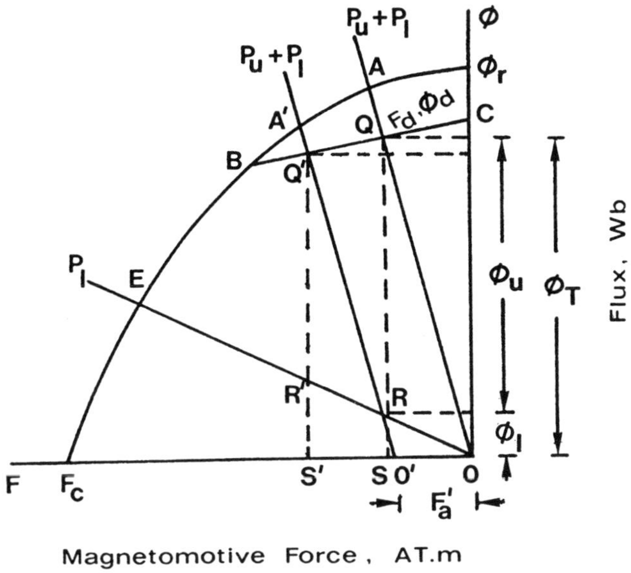

Fig. 4.11. A typical demagnetization curve.

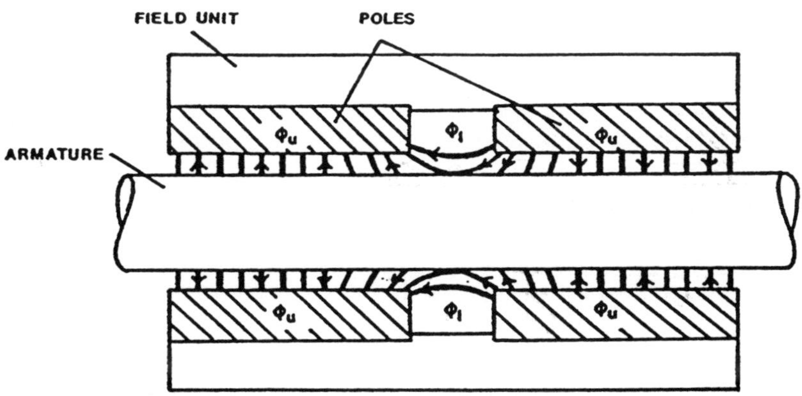

Fig. 4.12. The main and leakage flux paths.

Combining eqns (4.79) and (4.80) gives

$$\phi = H_u l_u P_u + H_l l_l P_l \,,$$

$$= -Hl(P_u + P_l) \,,$$

$$= F(P_u + P_l) \,. \tag{4.81}$$

Equation (4.81) gives the total unit permeance line, OA, through the origin which has a slope of $P_u + P_l$ (see Fig. 4.11). The intersection of this line

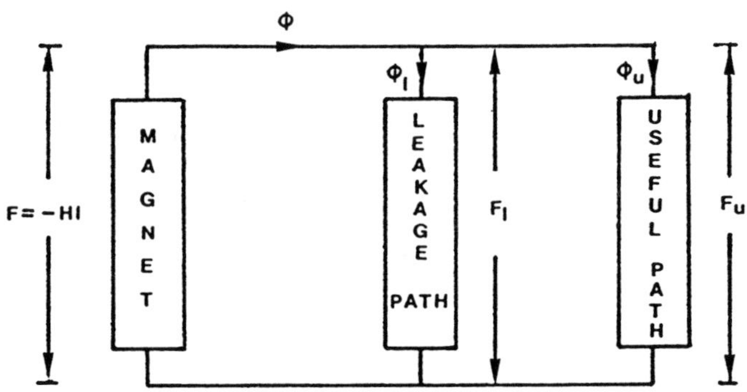

Fig. 4.13. An equivalent magnetic circuit in terms of m.m.f.

and the recoil line BC determines the working point, Q, of the magnet in the motor. The leakage unit permeance line, OE, is drawn with a slope of P_l. Now QR and RS give the useful and leakage fluxes, respectively, in the absence of an armature current. From Fig. 4.11,

$$F_u = F_l \,, \tag{4.82}$$

$$\frac{\phi_u}{F_u} = P_u \,, \tag{4.83}$$

$$\frac{\phi_l}{F_l} = P_l \,. \tag{4.84}$$

The reciprocals of the slopes of these lines give the useful and leakage reluctances $\mathfrak{R}_u$ and $\mathfrak{R}_l$ (see Fig. 4.14). The series term $\mathfrak{R}_o$ is given by

$$\Re_o = \frac{F_o}{\phi_0}, \tag{4.85}$$

where $\phi = \phi_0$ at $F = 0$.

When an armature current flows, an armature m.m.f. F_a is generated. Gould[34] has shown that F_a', the effective demagnetizing m.m.f., is a fraction of F_a. This demagnetizing m.m.f. divides between the magnet and the air gap in proportion to their respective reluctances, and it produces a demagnetizing field of H_a' on the magnet, where $F_a' = H_a' l_m$ and l_m is the magnet's length. Here,

$$-Hl_m = F_u + F_a' = \frac{\phi_u}{P_u} + F_a', \tag{4.86}$$

$$-Hl_m = \phi_l / P_l, \tag{4.87}$$

$$\phi = \phi_u + \phi_l. \tag{4.88}$$

Combining these three equations, we can write

$$\phi = -Hl_m(P_u + P_l)\left[1 + \frac{F_a' P_u}{Hl_m(P_u + P_l)}\right]. \tag{4.89}$$

Therefore, the total unit permeance line shifts by a factor of

$$\frac{F_a' P_u}{Hl_m(P_u + P_l)}.$$

The new line O'A' intersects the recoil line at Q' , giving a new operating point. It can be observed that the useful flux reduces to Q'R,' and the leakage flux increases to R'S' (see Fig. 4.11).

In the case of the double-armature d.c. linear motor[10] discussed in Chapter 2, and shown diagrammatically in Fig. 2.22, we can again adopt a linear magnetic circuit model as shown in Fig. 4.15. It is fairly obvious that the slider 'sees' through the air gap two armature reluctances connected in parallel. One of these reluctances $\Re_a$ is due to the magnetic path in one half of the motor, and the other reluctance $\Re_b$ is due to the magnetic path in the other half. Both of these reluctances depend on the position of the slider. The total reluctance is given by

$$\Re_{ab} = \left[\left(\frac{r+2x}{\mu\ l_1^2}+\frac{2l_0}{\mu\ A_0}\right)^{-1}+\left(\frac{r+2l_t-2x}{\mu\ l_1^2}+\frac{2l_0}{\mu_0 A_0}\right)^{-1}\right]^{-1}. \tag{4.90}$$

In eqn (4.90), r is a constant equal to $3L_c + 2(l_m + l_g)$, where l_m is the length of the side magnets, l_g is the air gap length, μ and μ_0 are the magnetic permeabilities of the core and air, respectively, l_1 is the active conductor length, l_0 is the length of each nonmagnetic gap along the armature core, A_0 is their cross section, and x is the distance of the centre of the slider from one end of the long static bars. From eqn (4.90) the expression for the magnet load line is

$$\frac{B_m}{H_m} = -\frac{\mu_0 l_m C_f L_f}{l_g + 2\mu_0 A_0 \Re_{ab}} \tag{4.91}$$

where C_f and L_f are the fringing-flux and leakage-flux factors, respectively, for the magnetic flux provided by the magnets; they are both dependent on the geometry of the motor. We can prove from eqns (4.90) and (4.91) that $\Re_{ab}$ has a parabolic behaviour with respect to x. Consequently, the magnetic flux provided by the magnets is a minimum at the centre and a maximum at the extremes of the armature core, as is the static thrust.

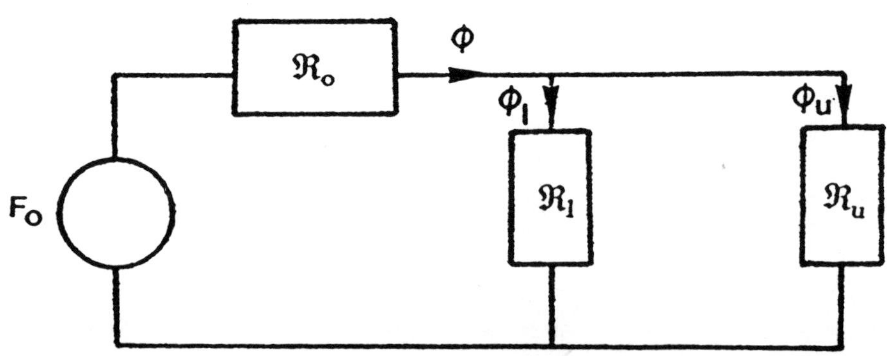

Fig. 4.14. An equivalent magnetic circuit in terms of reluctances.

For motors with square armatures and two back irons, all of the same length (see Fig.2.22), the effective field can be calculated by considering only one half of the motor, consisting of one permanent magnet and half the armature winding, due to the symmetry about the armature axis. Figure 4.16 indicates how the magnetic circuit with the winding unenergized can be represented by an electrical equivalent circuit with a constant-voltage source, resistances, and currents. The permanent magnet is the constant-voltage source in the equivalent circuit and it delivers an m.m.f. F_m. This m.m.f. has units of amperes and it is the product of the apparent coercive field strength H_c* (see Fig. 4.17), and the thickness of the magnet t_m, that is,

$$F_m = t_m H_c * \tag{4.92}$$

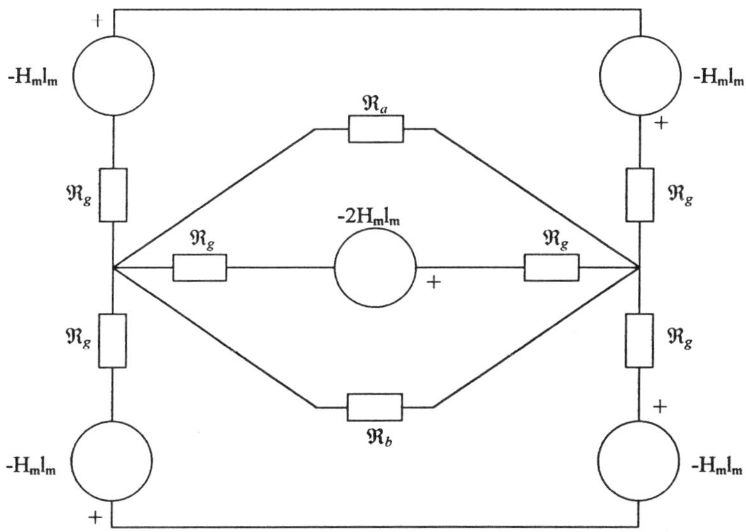

Fig. 4.15. A linear magnetic circuit for the double-armature motor.

The currents in the circuit are represented by the magnetic fluxes ϕ (in webers). The permeances Λ (in henrys) of the magnetic 'resistances' for a uniform field are

$$\Lambda = \frac{\mu_0 \mu_r A}{d}, \tag{4.93}$$

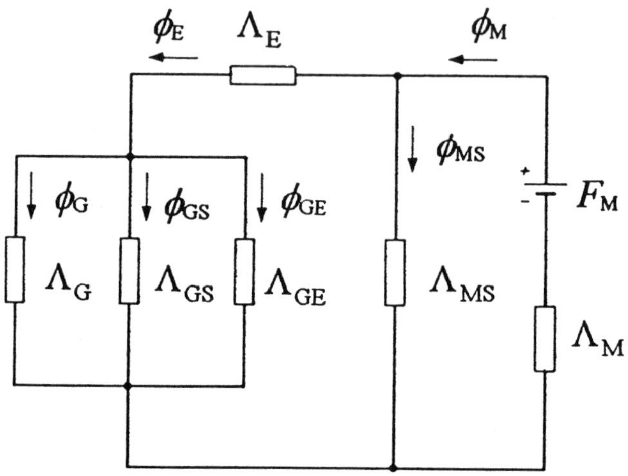

Fig. 4.16. Half of the linear magnetic circuit for the square-armature motor with two back cores.

where A is the area perpendicular to the direction of the lines of force, d is the length in the direction of the lines of force, μ_0 is the permeability of free space (in H m^{-1}), and μ_r is the relative permeability. Equation (4.93) can be used to calculate Λ_E, Λ_G, and Λ_M. The value of B_r / H_c* is taken to equal the product $\mu_o\mu_r$ when determining Λ_M. The other permeances in the equivalent circuit can be calculated from simple expressions that apply for idealized circular or elliptical lines of force.[41] Figure 4.18 shows an idealized geometry for the lines of force. The outer four lines of force are arcs of a circle; the other lines are parts of an ellipse. For this case, the permeance $d\Lambda$ of an element dx is given by[68]

$$d\Lambda = \mu_o\left[0.26 + \pi^{-1}\ln\left(\frac{\delta + 2a}{\delta}\right)\right]dx \ , \tag{4.94}$$

where δ is the length of the air gap and a is the half-width of the inner core and the full width of each of the outer cores. The motor being considered here has

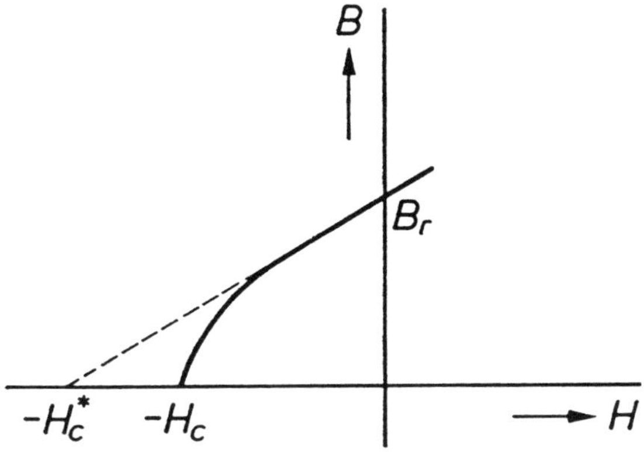

Fig. 4.17. The demagnetization curve for the permanent-magnet material described by Honds and Meyer[41] for their motor.

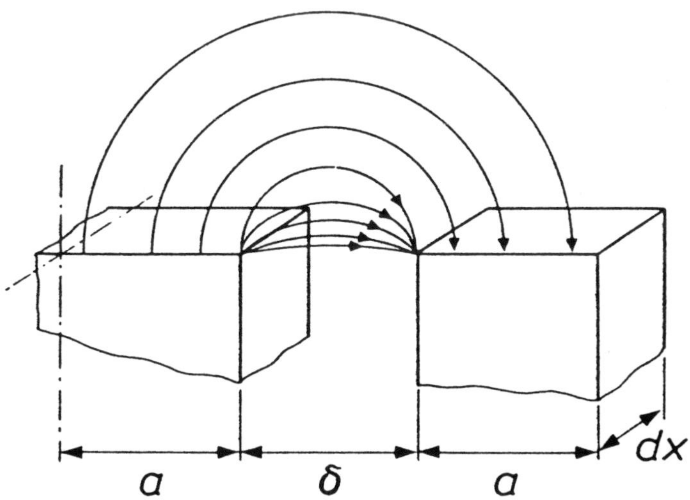

Fig. 4.18. An idealized geometry for the lines of force described by Honds and Meyer for their motor. Idealizations of this kind can be used for deriving simple expressions for the permeance.[68]

an inner-core width which is double the outer-core width.

We can now set up a number of equations in which the fluxes are the unknowns, and which can be solved for ϕ_E,

$$\phi_E = \phi_G + \phi_{GS} + \phi_{GE} , \tag{4.95}$$

$$\phi_M = \phi_E + \phi_{MS} , \tag{4.96}$$

$$\frac{\phi_{MS}}{\Lambda_{MS}} = \frac{\phi_E}{\Lambda_E} + \frac{\phi_G + \phi_{GS} + \phi_{GE}}{\Lambda_G + \Lambda_{GS} + \Lambda_{GE}} , \tag{4.97}$$

$$\frac{\phi_{GS}}{\Lambda_{MS}} = F_M - \frac{\phi_M}{\Lambda_M} , \tag{4.98}$$

where Λ_M, Λ_E, Λ_G, Λ_{MS}, Λ_{GS}, and Λ_{GE} are the permeances and ϕ_M, ϕ_E, ϕ_G, ϕ_{MS}, ϕ_{GS}, and ϕ_{GE} are the fluxes of the subfields around the magnets, the effective field in the air gap between the magnets and the inner core, the return field between two cores, the stray field around the magnets, the stray field at the two sides and the stray field at the two ends respectively.

If

$$\phi_E = A_M B_E \quad \text{and} \quad F_M \Lambda_M = B_r A_M \tag{4.99}$$

then the expression for the effective flux density can be written as

$$B_E = \frac{B_r}{\left[(\Lambda_G + \Lambda_{GS} + \Lambda_{GE})^{-1} + (\Lambda_E)^{-1}\right](\Lambda_{MS} + \Lambda_M) + 1} \tag{4.100}$$

This expression reduces to

$$B_E = \frac{B_r}{(1/\Lambda_G + 1/\Lambda_E)\Lambda_M + 1} \tag{4.101}$$

if the effect of stray fields is neglected, and we have $\Lambda_{GS} = \Lambda_{GE} = \Lambda_{MS} = 0$.

The theories discussed in this chapter will enable readers to understand the working principles of various types of dc linear motor and to design motors. After the preliminary design, various methods, which are described in Chapter 7, can be used to analyse the performance. This analysis will help readers to produce an optimal design.

5

SUPPORTS AND GUIDANCE

The rotary equivalents of linear motors are much easier to support and guide. As the travelling distance of the linear motor increases, the problem of aligning the armature is accentuated since the travelling distance determines the length of the stationary unit. So a considerable air gap must be provided in order to avoid any unwanted contacts due to any imbalance in the system.

5.1 Wheels and bearings

A plan of a basic system is shown in Fig. 5.1. The core of the armature, which is stationary in this type of linear motor, was made by joining five sections of 3 m long, circular, mild-steel bars. The armature is mounted at a height of 10 cm from the base, a wooden plinth, by Tufnol supports ($10\times5\times1$ cm^3) placed at 1 m intervals. The field system is mounted inside the carriage, which is the moving part of the linear motor (see Fig. 5.2). The carriage is also made from Tufnol, and it has four nylon wheels which run on a pair of rails which are fixed on the wooden plinth. The distance between the two parallel rails is constant; however, due to the long travelling distance the minimum air gap between the field and the armature was chosen as 3 mm, which is considerably larger than the air gap for a rotary motor.

The support and guidance for the moving unit can also be accomplished by a rail which is the same length as the armature core and is placed parallel to it, over the motor (see Fig. 5.3). The carriage has two supports, one fixed at the front and the other at the rear of the carriage, each consisting of two nylon wheels which run on this overhead rail. In order to maintain a balanced air gap on either side of the armature, and to prevent wobble, bounce, and vibration of the field system, two nylon rollers with an adjustable mechanism have also been fitted to each of the end plates of the carriage. These rollers run on the armature winding.

The powerful neodymium–iron–boron magnets housed in the moving unit produce a force of attraction between the poles and the armature on either side of the axis. The net force flattens the surface of the nylon after some usage, hence it increases the friction between the armature windings and the carriage. This friction, in addition to the friction between the nylon wheels and

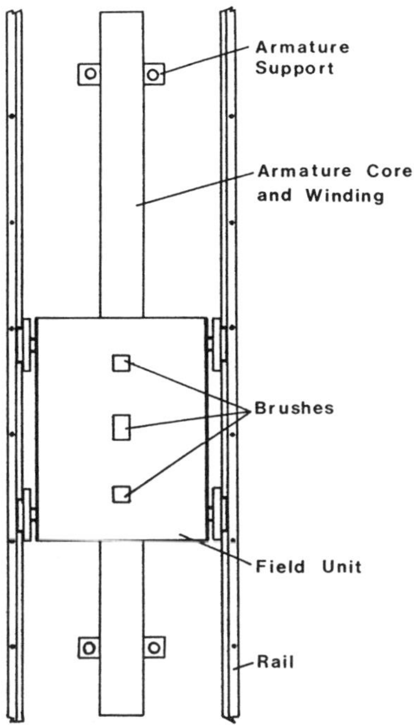

Fig. 5.1. A plan of a basic d.c. linear motor.

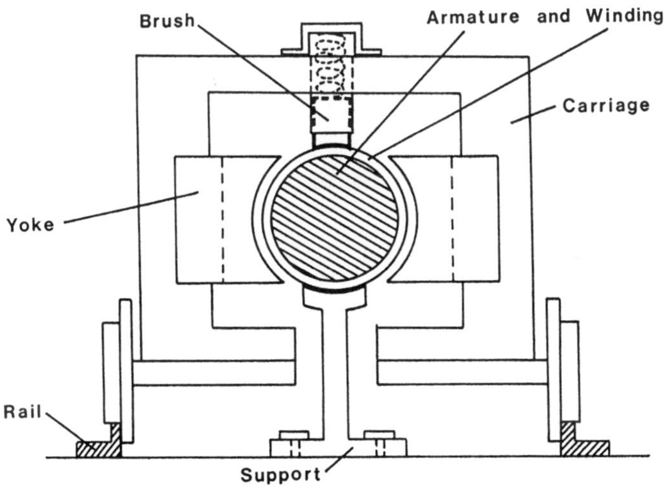

Fig. 5.2. An elevation view of a basic d.c. linear motor.

the rail, causes many mechanical problems. In order to overcome these difficulties a new method of suspension of the carriage has been implemented. A slotted linear bearing with its housing has been fixed to the carriage framework on either side of the armature. These bearings slide on two perfectly parallel, 25 mm diameter, mild-steel rods, each of which is the same

Fig. 5.3. A linear motor with an overhead support.

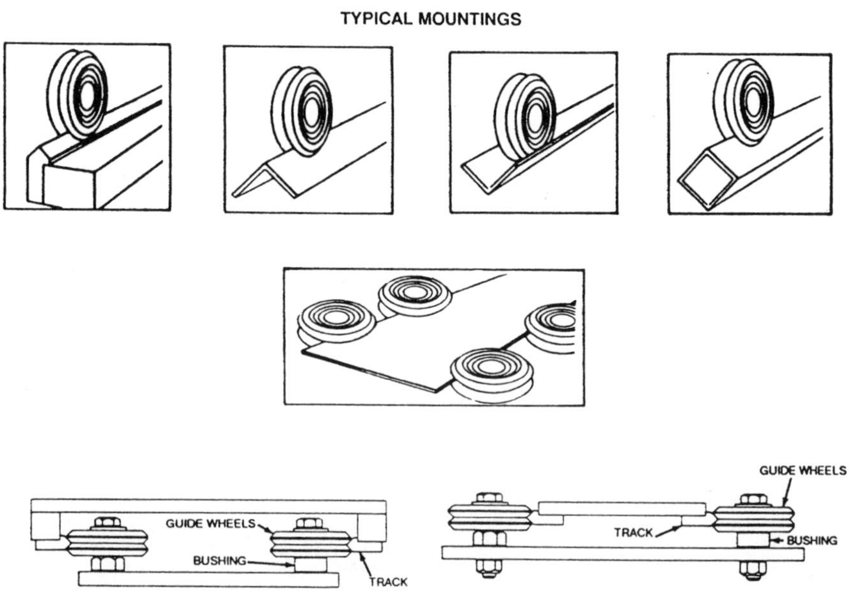

Fig. 5.4. Linear guided systems made by Techno.

length as the armature, and is supported by six U-shaped structures. The height of the two rods from the base, and the distances between them and the armature core, can be adjusted with the aid of several threaded rods (studding).

The armature core is thus aligned to the field system, hence the air-gap length is more uniform. Any possibility of friction is consequently reduced. Several commercially available guided systems are shown in Fig. 5.4.

5.2 U-spring supports

Different types of d.c. linear motor are now being developed for use in various fields. It is difficult to use moving machines in sites like clean rooms, because the fine dust or powder which is produced by friction in the moving parts of machines enters the atmosphere, and hence damages, for example, the materials or circuitries being worked upon. In order to overcome this problem,

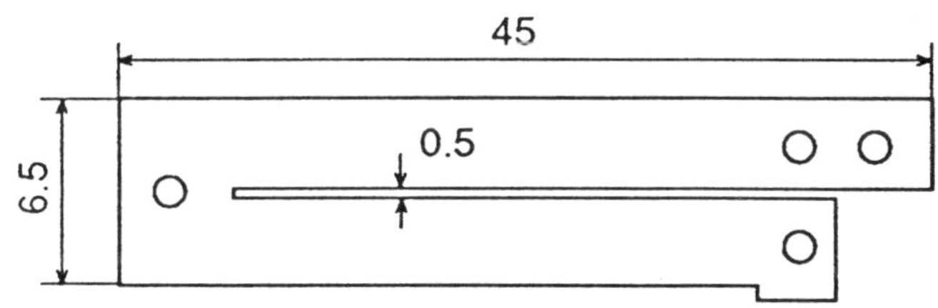

Fig. 5.5. A U-shaped spring 0.15 mm thick (all units are in millimetres).

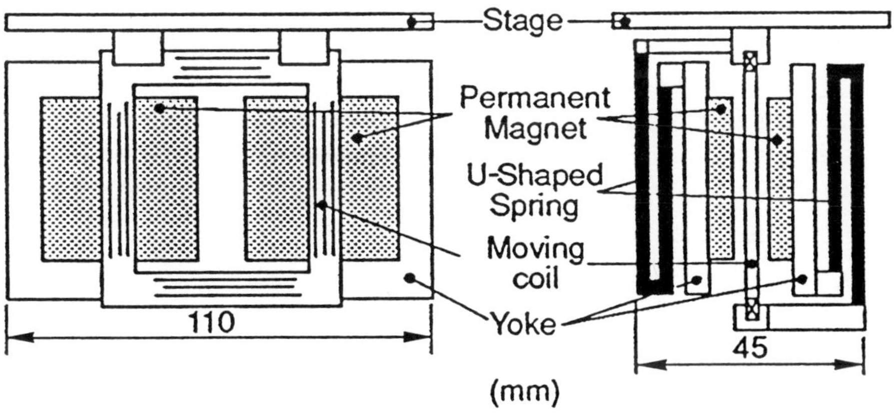

Fig. 5.6. Side and front views of the motor with U-shaped springs.

some schemes have been investigated; however, the control of these systems is not easy, and the facilities are expensive. A new supporting mechanism, which is of quite simple construction, inexpensive, and easily controllable, is described[57] in this section.

In a motor which has a stroke of ±8 mm, four frictionless U-shaped springs (see Fig. 5.5) were used as a new type of supporting mechanism for the mover, which is a coil. This supporting mechanism not only holds the mover but is also frictionless, and it can be moved without any upward or downward displacement. The yokes and the permanent magnets comprise the stationary part. Two pairs of magnets are arranged as shown in Fig. 5.6, so that the flux density in the air gap between the yoke and the coil is uniform. The position of this motor is detected by a linear, variable differential transformer with a nonlinearity of only 1 per cent.

5.2.1 *The reactive force of U-shaped springs*

The reactive force of the springs is proportional to their displacement. The force due to the four springs attached to the motor is shown in Fig. 5.7(a). Observe that the force characteristic is linear within the maximum stroke length. The spring coefficient was calculated as 95 N m^{-1}.

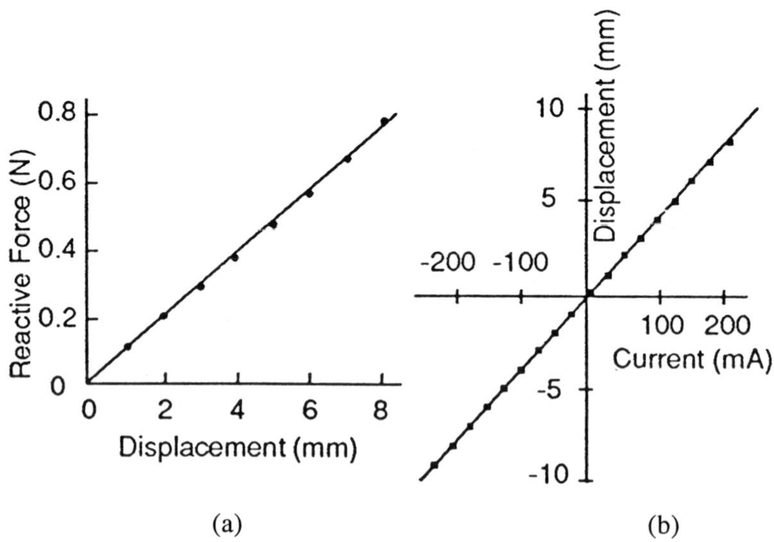

Fig. 5.7. Motor characteristics.

The displacement-versus-current characteristic of the motor is shown in Fig. 5.7(b). It can be seen that the displacement of the mover, with the current, is again linear. The thrust coefficient was estimated as 3.9 $N A^{-1}$ from the results shown in Fig. 5.7.

5.2.2 *The controller*

Several controllers for driving the motor with U-spring supports as described above have been designed and tested. The characteristic of the controllers has been gradually improved, and finally a controller has been designed to produce the optimal result. Figure 5.8 shows the block diagram of its circuit. The proportional, integral and derivative (PID) controller is in the main loop, and the error signal which is obtained from the desired signal and the position detector is an input to it. The frequency range of the supporting springs and the PID controller are eliminated by a fourth-order low-pass filter placed in the subfeedback loop, followed by the proportional and derivative (PD) circuit. The PD controller compensates for the delay of phase due to the active filter. The result for this controller is shown in Fig. 5.9. The maximum following frequency is about 15 Hz, which is the frequency of the PID controller. Only the peak-frequency band which appears in the PID circuit of the controller is eliminated. The difference between the phases is about −40°. This controller improves both the gain and the phase difference.

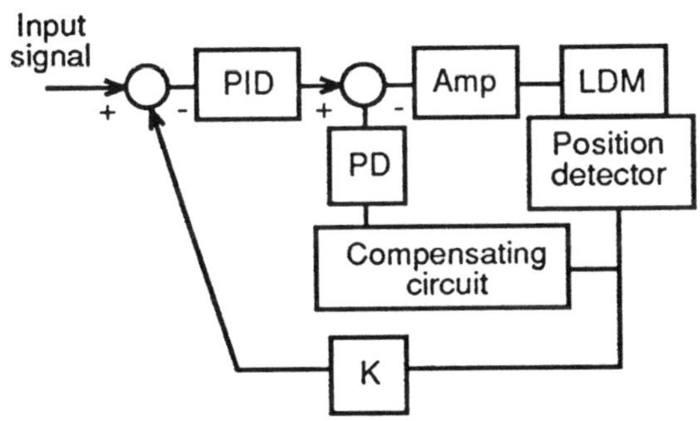

Fig. 5.8. A block diagram of the PID control and the compensating circuit.

5.3 Magnetic levitation systems

From the practical point of view, the bearings and the U-shaped springs supporting the moving unit of a linear motor affectits operating characteristics.

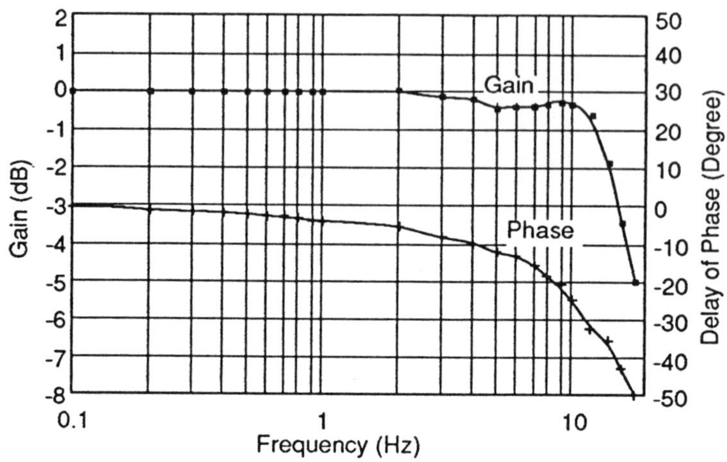

Fig. 5.9. Frequency characteristics of the PID control and the compensating circuit.

Compressed-air suspension and magnetic levitation are examples of the effort to effect contactless suspension.

Magnetic levitation of the moving units of linear motors is receiving increased attention because of the development of magnetically levitated railway systems. Considerable research and development has been carried out, and still continues, on rail-transport systems using magnetic levitation for support. Fruechte *et al.*[29] have given design details in a recent paper on a test vehicle which was constructed to obtain operating experience with a low-speed system intended for personal, rapid-transit applications. Levitation and lateral guidance are provided by four pairs of U-shaped, d.c.-controlled electromagnets which attract the vehicle towards a U-shaped steel rail. Each of the four pairs of magnets (one pair at each vehicle corner) is individually controlled to maintain a certain vertical air gap and to provide lateral support and damping. Vertical support of the vehicle weight by an attractive constant-current magnet is inherently unstable since the attractive force decreases as the air gap increases, and thus the slightest disturbance from equilibrium causes the vehicle either to fall or to clamp to the rail. In order to stabilize this motion a feedback system is employed. The two magnets of each pair are laterally offset by equal amounts (see Fig. 5.10); thus for equal currents the lateral forces cancel each other out. For lateral-motion control, as one current increases the other decreases by the same amount; this gives the required lateral force but does not change the vertical force.

Magnetic forces are often calculated with the aid of the virtual-work method. Figure 5.11 shows a model[27] for vertical motion. We assume that the current is constant at i_{mo}, the air gap at l_{go}, and the coil inductance at L_o. The inductance changes with the air-gap length; therefore L is a function of l_g. This function can either be estimated from magnetic-circuit theory or it can be obtained experimentally. The energy stored in the inductor under these conditions is

$$E = \tfrac{1}{2} L(l_g) i^2_{mo} \tag{5.1}$$

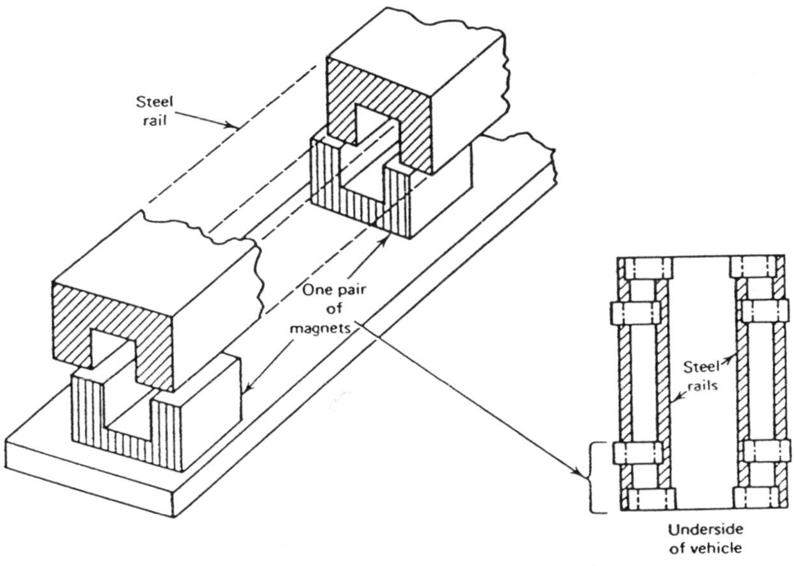

Fig. 5.10. A magnetic levitation system (courtesy of E. O. Doebelin[27]).

Now assume that an infinitesimal displacement, dl_g, occurs while the current is held constant. If the vertical magnetic force is f_{ml}, the mechanical work done is given by $f_{ml}\, dl_g$. This work is equal to the change in the energy stored in the inductor:

$$\text{change in stored energy} = \tfrac{1}{2} L(l_g) i^2_{mo} - \tfrac{1}{2} L(l_g + dl_g) i^2_{mo} \tag{5.2}$$

or

$$\text{change in stored energy} = -\tfrac{1}{2}\,(dL) i^2_{mo} \tag{5.3}$$

or

$$f_{ml}\, dl_g = -\tfrac{1}{2}\,(dL) i^2_{mo} \,. \tag{5.4}$$

therefore,

$$f_{ml} = -\frac{i^2_{mo}}{2} \frac{\partial L}{\partial g}\bigg|_{l_g = l_{go}} \tag{5.5}$$

A partial derivative for the rate of change of L with respect to l_g has been used because L also depends on horizontal motions. We may write, for small perturbations (denoted by a subscript p) about an operating point:

$$f_{mlp} = \frac{\partial f_{ml}}{\partial i_m} i_{mp} + \frac{\partial f_{ml}}{\partial l_g} l_{gp} \,,$$

$$= -i_{mo} \frac{\partial L}{\partial l_g} i_{mp} - \frac{i^2_{mo}}{2} \frac{\partial^2 L}{\partial l_g^2} l_{gp} \tag{5.6}$$

Now, in the electrical circuit, the induced voltage, e, in a coil of cross-sectional area, A, exposed to uniform magnetic flux density, B, is:

$$e = \frac{d\psi}{dt} = \frac{d}{dt}(BA) \tag{5.7}$$

where ψ is the flux linkage; that is,

$$\psi = Li \,. \tag{5.8}$$

The current–voltage relation for the inductance L is given by

$$e_L = \frac{d\psi}{dt} = \frac{d}{dt}(Li) = L\frac{di}{dt} + i\frac{dL}{dt} \tag{5.9}$$

The output voltage of an electronic power amplifier is denoted by e_i in Fig. 5.11, and R and L denote the total resistive and inductive effects of the amplifier output circuit and the coil. We can obtain the following expression by using Kirchhoff's law:

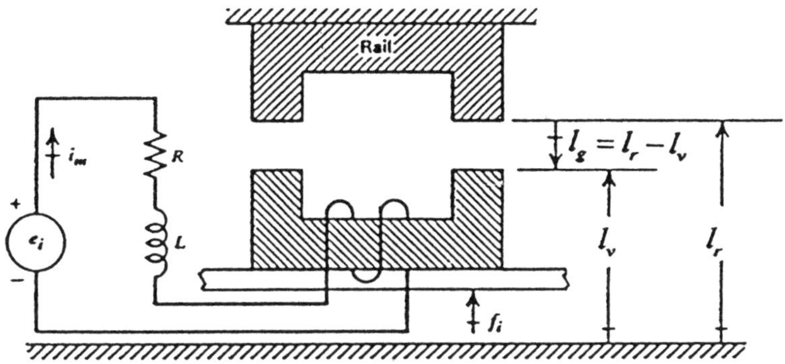

Fig. 5.11. A model for vertical motion (courtesy of E. O. Doebelin[27]).

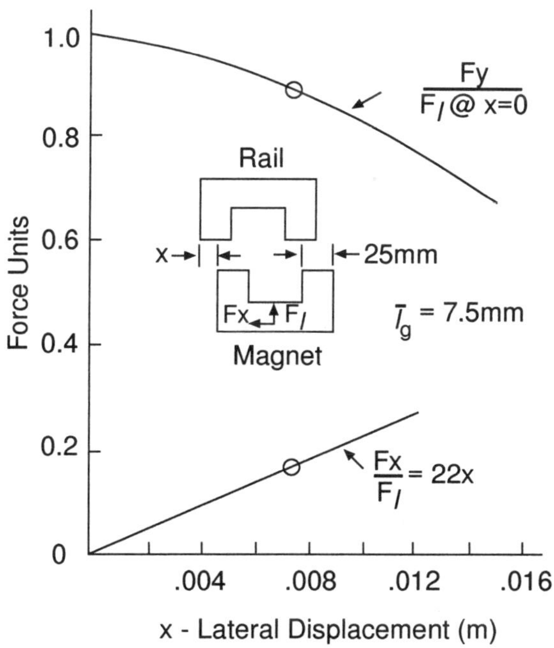

Fig. 5.12. The measured magnetic force characteristics (courtesy of E. O. Doebelin[27]).

$$e_{\rm ip} - i_{\rm mp} R - L \frac{di_{\rm mp}}{dt} - i_{\rm mo} \frac{\partial L}{\partial l_{\rm g}} \frac{dl_{\rm gp}}{dt} = 0 \qquad (5.10)$$

We can then find an expression for the vertical gap, $l_{\rm gp}$, by eliminating $i_{\rm mp}$, and combining the mechanical and electrical equations, eqns (5.6) and (5.10), respectively.

With the aid of the experimentally obtained magnetic-force characteristic illustrated in Fig. 5.12, Doebelin[27] has shown that the vertical force decreases as the lateral displacement increases, and that the ratio of the lateral to the vertical force varies linearly with displacement.

Further treatment of lateral guidance is beyond the scope of this book. Readers are advised to refer to the book by Doebelin[27] which comprehensively describes the levitation of a transport system.

5.4 A suspended direct-current linear motor

The moving-magnet-type linear d.c. motor with a short stroke developed by Matsumura *et al.*[58] is discussed in this section. The moving unit of this motor is supported in a completely contactless way. Figure 5.13 shows the part which produces the thrust. Two permanent magnets, each magnetized in the radial direction, but with opposite polarities, are placed in the moving unit. The magnetic paths are represented by solid and dashed lines. When the air gap between the permanent magnets and the yoke is sufficiently small, the magnetic paths shown by the solid lines become more dominant, reducing the leakage flux and the magnetic interaction with the levitated part. The armature current interacts with the radial flux to produce a thrust.

A cross-section of the complete motor is shown in Fig. 5.14. The moving unit is supported by a magnetic bearing at either end. The length of the air gap at a bearing is measured by an eddy-current-type position sensor. In order to detect the axial displacement of the moving unit, a linear encoder has been placed at the centre of the moving unit.

Compact and light permanent magnets have been used to keep the weight of the moving unit as small as possible. Matsumura *et al.*[58] have used a rare-earth cobalt magnet for this purpose, and they chose a permeance coefficient between 1.5 and 2.0 in order to let the flux density at the working point be as large as possible.

5.4.1 *Static characteristics*

The static thrust is one of the most important characteristics of linear motors,

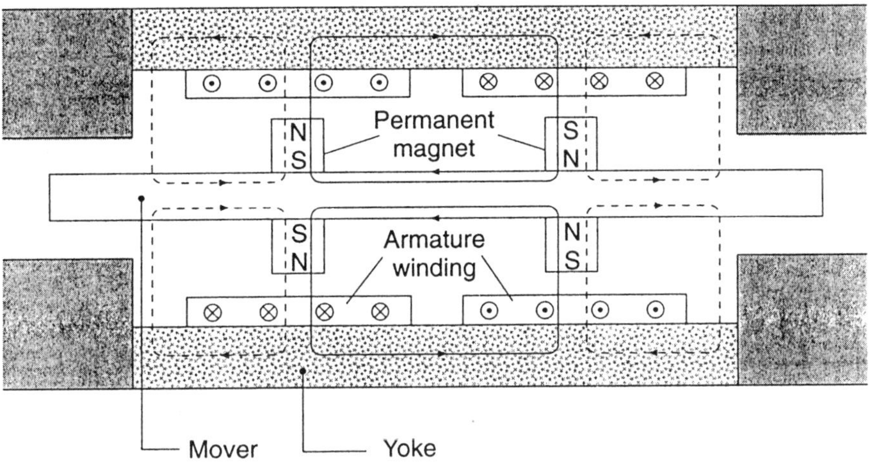

Fig. 5.13. A magnetic circuit with permanent magnets.

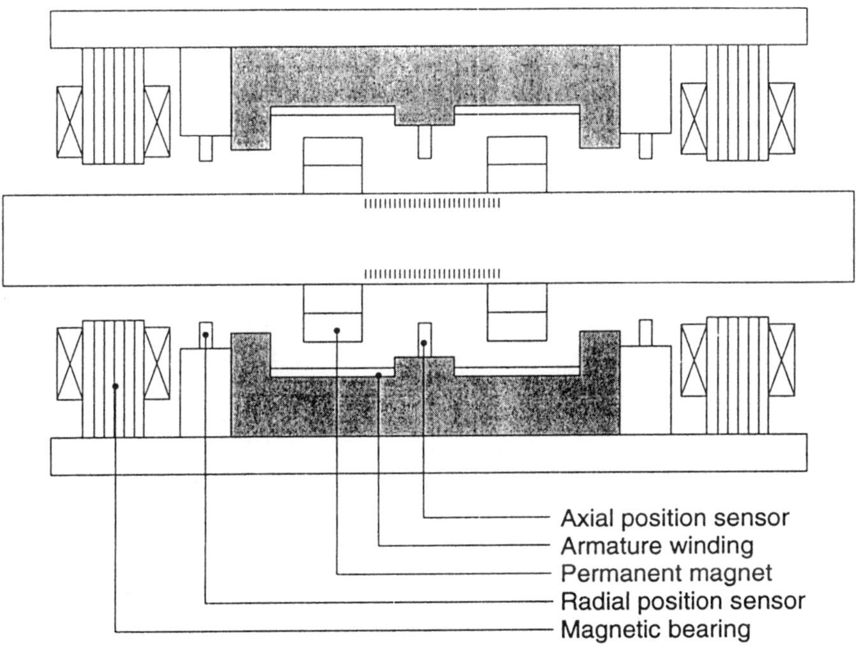

Fig. 5.14. A cross section of a suspended d.c. linear motor.

and therefore, in this case, it was measured for various values of the armature current and displacements of the moving unit. Figure 5.15 shows the experimental results, which show that an armature current of 1 A produces a static thrust of 3.7 N. These characteristics also show that the thrust is proportional to the armature current if $x_s = 16$ mm, but it becomes saturated by 12 per cent when the armature current is 2 A and $x_s = -16$ mm. This means that the right coil produces a greater thrust than the left coil at armature currents which are larger than 1 A. This could be affected by the magnetic bearing. Various experimental results are shown in Table 5.1.

5.4.2 *Driving characteristics*

In order to determine the suspension characteristics of the moving unit of this motor under reciprocating motion, the axial displacement, x_s, of the moving

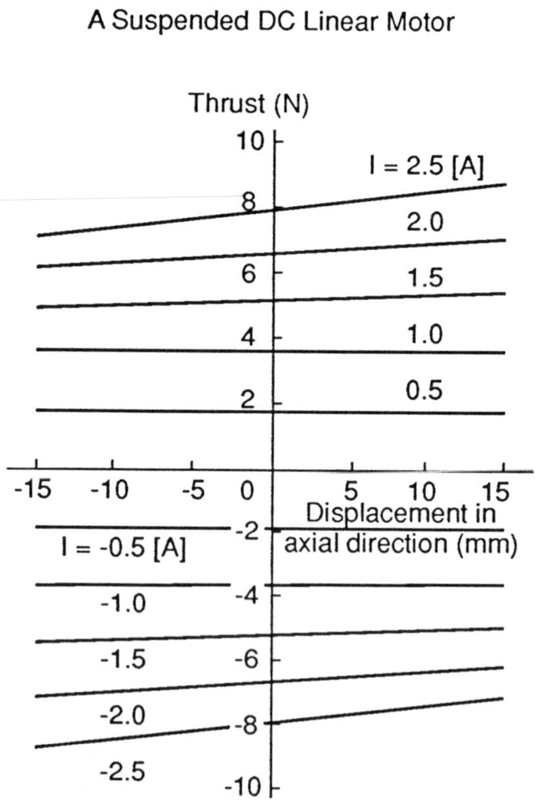

Fig. 5.15. Thrust-versus-armature-current characteristics.

Table 5.1 Characteristics of Matsumara's motor

Variable	Value	Unit
Stroke	0.045	m
Static thrust	5.5	N
Input power	0.021	kW
Thrust/input power	262	N kW^{-1}
Thrust/mass	38.5	N kg^{-1}
Thrust/volume	91.7	N m^{-3}
Thrust coefficient	3.7	N A^{-1}
Mover mass	0.255	kg
Flux density	0.3	T
Number of turns of coil	140	turns
Coil resistance	9.1	Ω
Coil inductance	0.01	H
Temperature rise	0.54	K W^{-1}

unit was measured with the help of a linear encoder which had a resolving power of 0.5 mm. The armature current was generated by a microcomputer, and its value oscillated between $-I$ and $+I$. Figure 5.16 shows the variation of the vertical-gap length with time. The curves indicate that the gap length was kept within 0.35 mm, and therefore the moving unit was suspended at a fairly precise position. This small variation in the air-gap length may be caused by irregularities in the assembly and by external disturbances.

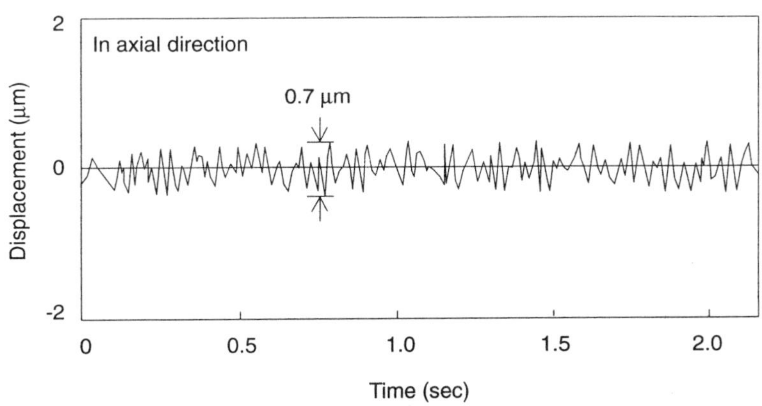

Fig. 5.16. The suspension state of the core in reciprocating motion for a stroke of 16 mm and a coil current of 44 mA.

5.4.3 *Suspension friction force*

The moving unit is freely suspended, hence there is no friction due to mechanical contact. However, there is a damping force due to eddy currents and magnetic hysteresis. The motion of the moving unit can, therefore, be described by

$$m\frac{dv}{dt} = K_f i - \left(Dv + K\, x_s + f_c\right), \tag{5.11}$$

where, $v = dx_s/dt$, K_f is the thrust coefficient, m is the mass of the moving unit, D is the damping force coefficient, and $f_c = F_c\ \mathrm{sign}(v)$.

The inductance of the coil and the counter electromotive force induced in the coil have little effect since the power amplifier works as a constant-current source. We can determine K_f and F_c from a knowledge of the armature current at the instant when the moving unit begins to move. For this motor the values are $4\times10^{-3}\ \mathrm{NA^{-1}}$ and 0.03 N.

The damping- force coefficient has been found to be $0.85\ \mathrm{Nm^{-1}s}$ by plotting the phase-plane trajectory under the reciprocating system.[58]

The discovery of high-energy product neodymium–iron–boron permanent magnets has made it possible to build a reliable low-friction suspension system for a linear motor. The main advantages of this system compared to other forms of suspension, are:

1. no power is required for lift after initial magnetization of magnets,
2. the cost is low,
3. there is no environmental pollution and
4. no servicing is required.

Although we may conclude from this chapter that magnetic levitation is the best form of support and guidance, all known manufacturers of d.c. linear motors are presently using linear bearings and V-section wheels in their products.

6

CONTROL METHODS

This chapter describes a system which was developed to control a 3 m long, d.c. linear motor with 64 armature winding sections. Several position-sensing devices which produce feedback signals for the control systems are also discussed. In addition, a software package which was developed to control the 3 m long motor is described.

6.1 Sensors

A sensor is a necessity in order to indicate the position and speed of a motor with an acceptable accuracy, and hence to control both the position and speed of the motor.[76] The Cranfield Unit for Precision Engineering has demonstrated that a transducer with a 0.01 micrometre resolution can repeatedly reposition a brush-type d.c. linear motor to within ±0.01 μm. The positional accuracy of the system was limited by the position transducer, and not by the linear motor.

The sensor has to be inexpensive and it should be highly unaffected by interfering signals. Its performance should be independent of temperature changes and environmental conditions. In general, the position of the motor is sensed and the signal from the sensor is then sent to a microcomputer to control the position and speed of the motor.

One of the following sensors can be used for detecting the position.

1. *Ultrasonic sensors.* An ultrasonic source transmits a pulse, and a digital counter is used to calculate the time taken for the ultrasonic pulse to travel from the transmitter to the motor and to reflect back to a receiver which is placed by the source.

2. *Optical sensors.* A light source and a detector are mounted on the carriage. When the carriage moves, a series of pulses are produced by obstacles placed at equal intervals along the track; the number of pulses is proportional to the travelling distance of the carriage.

3. *Magnetic sensors.* A Hall-effect vane switch, which detects the presence of a ferrous-metal vane passing through the gap between the sensor and a magnet, produces a bounce-free switched output.

This system is more applicable in dusty or high-ambient-light environments where an optical method would be unsuitable. For a permanent-magnet d.c. linear motor, however, it is undesirable to use such sensors because of the influence of the motor's permanent magnets on the sensor. Magnetic shielding will be required in this case.

4. *Linear variable differential transformers (LVDTs).* Linear variable differential transformers are one of the most common displacement transducers. They consist of three coils, one for energization and two for pickup. The coils are wound on a cylindrical former with a central magnetic core. The primary winding is driven by a sinusoidal voltage, and the amplitude and phase of the output signal are directly proportional to the position of the core with respect to the pick-up coils. The output signal is demodulated by a phase-sensitive detector.

Linear variable differential transformers have high resolution, accuracy, and linearity; and they have no wear problems. However, due to their limited range of stroke length, they are unsuitable for very long motors.

5. *Resistance sensors.* The operation of resistance sensors depends on Ohm's law. By applying a constant current through a 24 s.w.g.(standard wire gauge) wire of high resistivity (Ni–Cr) placed along the track, with a sliding mechanism mounted on the carriage, as shown in Fig. 6.1, a voltage from another 24 s.w.g. wire of comparatively low resistivity (Cu) can be detected. This voltage represents the position of the motor. When the motor moves along the track, the detected voltage is proportional to the displacement. This method is inherently inaccurate, but it is still the simplest and cheapest method.

In the following sections we will discuss the application of three of the five sensors described above.

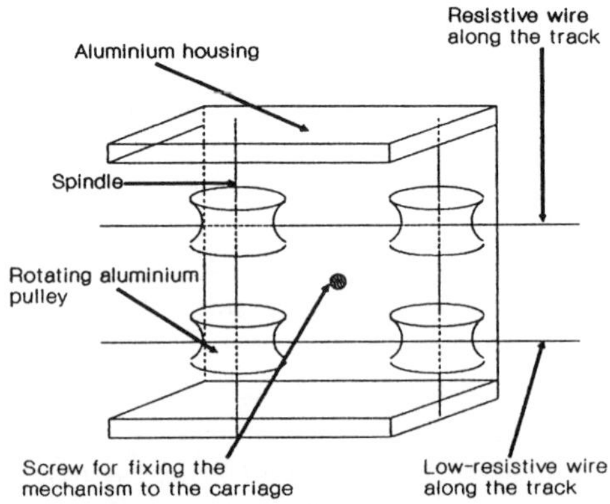

Fig. 6.1. A sliding mechanism.

6.1.1 *Ultrasonic transducers*

Ultrasonic transducers convert an electrical signal to sound energy, and vice versa. One transducer, acting as a transmitter, sends out a short ultrasonic pulse, the echo of which is received by another transducer which acts as a receiver. The time between transmitting and receiving the signal is proportional to the object's distance. Since we know the speed of sound in the transmission medium, we can calculate the distance.

The power of the transmitted signal decreases with $1/d^2$, where d is the distance, and the attenuation increases with the frequency. The frequency should not be too low, otherwise a reduction in the directivity of the beam is experienced, and an audible sound is emitted. The optimum frequency[28] for applications in air is in the region of 40 kHz.

A block diagram of a typical ultrasonic-sensor circuit is shown in Fig. 6.2. This comprises a transmitter circuit, a receiver circuit, an eight-bit counter, an eight-bit latch, and a control circuit. The transmitter and receiver are fixed side by side at one end of the track while a reflecting object, which covers the area of both transducers, is mounted on the motor carriage. The distance travelled by a motor can be calculated from the time between transmitting the signal and receiving the echo.

The transmitter circuit is usually an astable multivibrator. It is built using a quad two-input NOR gate and external components, as shown in Fig. 6.3, to produce 40 kHz ultrasonic pulses with a 12 V peak amplitude. The pulses are used to drive an ultrasonic transmitter. The frequency of oscillation can be adjusted by varying the adjustable resistor R_1 for a maximum sensitivity of 106 dB.

A sensitive receiver circuit may consist of a two-stage preamplifier and a comparator circuit. Its sensitivity can be changed by adjusting the potentiometer supply voltage at the noninverting input of the comparator circuit, as shown in Fig. 6.4. The ultrasonic pulse received by the circuit drives the output of the operational amplifier to the high potential (+12 V), and as a result the potential divider ($R_1 + R_2$) switches on the final-stage transistor, and the output level becomes high.

Both the transmitter and the receiver are connected to their respective circuits via coaxial cables, which act as noise suppressors. In order to measure the time between transmitting and receiving the ultrasonic pulses, we used an eight-bit counter which is fed from a pulse source of known frequency to count the number of pulses during the transmission period. A timer circuit is used for the pulse source. The period of time can be adjusted by simply varying a resistor. The output of this counter is connected to the input of the eight-bit latch circuit.

Operation of the feedback circuit. As the first monostable mutivibrator in the sensor circuit (see Fig. 6.2) is fired by the pulse from the output port of the

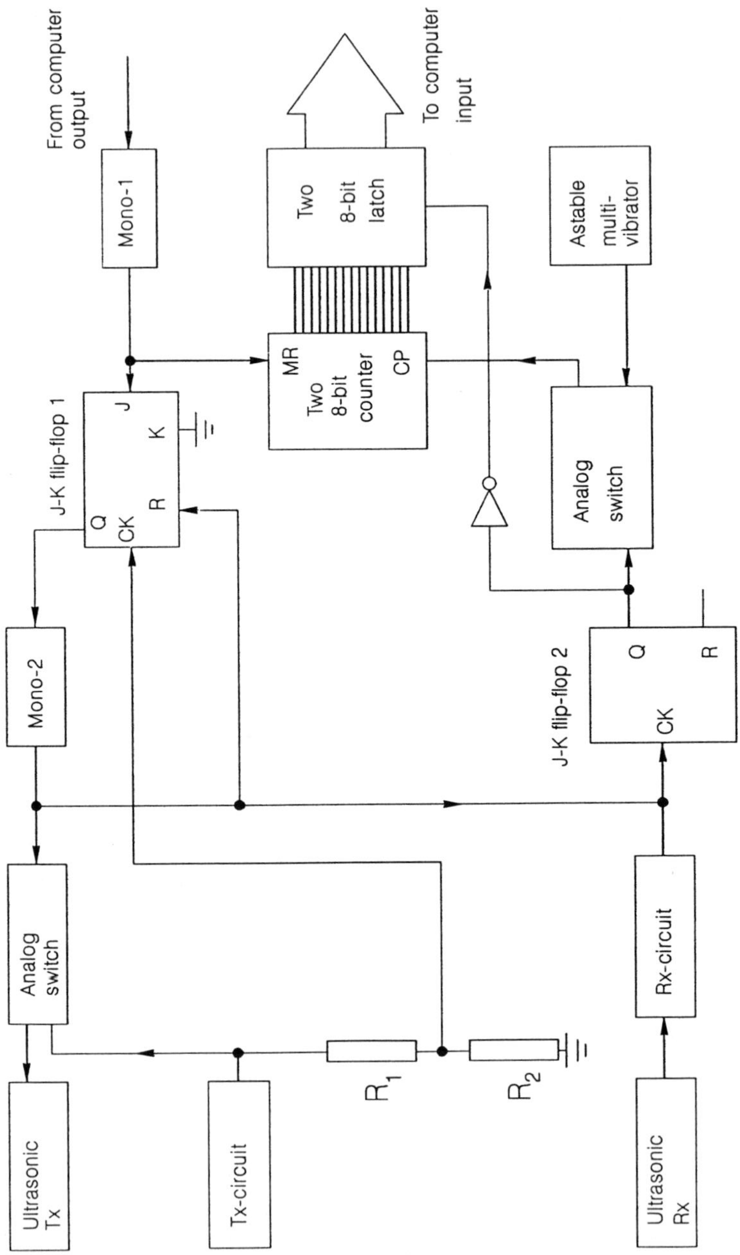

Fig. 6.2. A block diagram of an ultrasonic-sensor circuit.

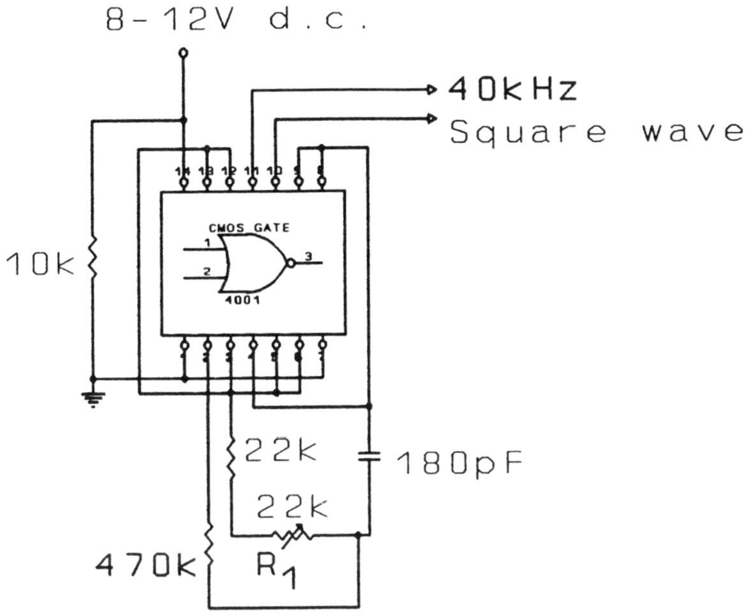

Fig. 6.3. A transmitter circuit for ultrasonic sensing.

computer, it produces a pulse with a 50 μs width, causing the counter to reset. Simultaneously, this pulse causes the Q output of the first *JK* flip-flop to be set to 5 V in association with the rising edge of a pulse from the transmitting circuit. The duration of this pulse determines the period of time for which the counter is kept reset. Since the pulse train has a 12 V peak amplitude, a potential divider ($R_1 + R_2$) was used.

The second monostable multivibrator is fired and this generates a pulse with a 60 μs width, and this turns on the analog switch causing the transmission of an ultrasonic pulse; at the same time, this pulse sets the second *JK* flip-flop and it resets the first *JK* flip-flop. By setting the second *JK* flip-flop, the second analog switch is turned on, allowing the astable pulse source to feed the counter.

Once the ultrasonic transmitted pulse has been received by the receiving circuit, the second *JK* flip-flop is reset, which stops the astable pulse source from feeding the counter. The Q output of the second *JK* flip-flop is simultaneously triggered after being inverted, allowing the value of the counter held in the latch registers to feed the input port of the computer.

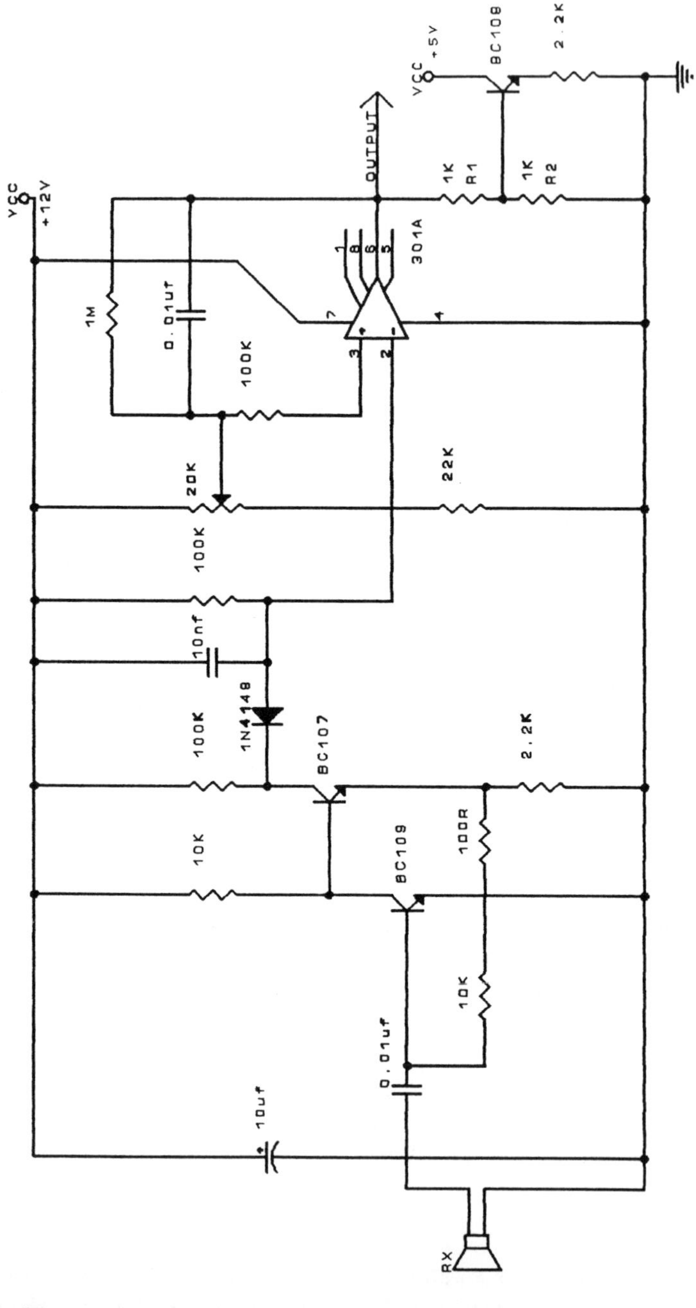

Fig. 6.4. The receiver circuit for the ultrasonic sensing.

Implementation of this ultrasonic technique so that the feedback circuit can sense the motor position is a reliable and accurate method in applications requiring a 2 mm resolution. It offers a noncontact method of sensing the position, and thus of reducing the overall friction of the system. The ultrasonic sensors do not function adequately at distances of less than 30 cm, so they must be set at least 30 cm away from the starting point of the motor.

The beam pattern of ultrasonic transducers offers a wide directional angle. When the moving unit of the motor, where the reflector is usually mounted, is far away from the transmitting end, the receiving circuit receives the transmitted pulse that is reflected back from the reflector in addition to the pulses reflected from other objects, such as the frames holding the tracks. This causes an error signal; therefore, the feedback circuit needs a very acutely angled beam system. In order to obtain a beam angle of less than 10° with these transducers, an array may be constructed. The design of such an array is complicated and consequently expensive; however, such arrays are extremely reliable when beam angles as small as 1° need to be implemented.

An alternative method for sensing the position of the motor is given in the next section.

6.1.2 *Optical sensors*

A slotted opto-switch has an infra-red source and a detector; the source illuminates the detector directly, and the slot allows an object, such as a rotating disc with vanes or holes, to interrupt the light beam.[8] Thus, pulses are generated whose number depends on the number of light interruptions.

The source is a gallium-arsenide infrared light-emitting diode, and the detector generally comprises a Schmitt trigger, a voltage regulator, a differential amplifier, and a photodiode.

This type of sensor can be used for a linear motor, as shown in Fig. 6.5. A slotted opto-switch is attached to the moving unit. A toothed metal strip, similar to a comb, is used instead of a disc. It is laid out along the guiding track, and the switch glides with the strip inside its slot and thus generates a train of pulses.

We can basically understand the optical-feedback circuit from the block diagram shown in Fig. 6.6. This circuit continually indicates the position of the motor, enabling accurate positioning of the motor and appropriate winding energization. It consists of two circuits; the first is mounted on the moving unit, and the second is at the receiving end of the system.

The moving part of the circuit mainly comprises a slotted opto-switch, a modulator, an amplifier, and a high-power light-emitting diode. The stationary part of the circuit comprises among others a high-power photodiode and an up–down counter circuit.

Fig. 6.5. A toothed metal strip used to generate trains of pulses.

Any interruption of the infrared light beam emitted from the source causes the modulator to reset. The output of the modulator is amplified and fed to the base of the output power transistor; this turns the final-stage transistor on. The high-power emitter diode is a high-radiant-intensity GaAs infrared-emitting diode which is placed, forward biased, in the collector circuit of the final-stage transistor. Once the transistor is driven on, the emitter starts to radiate power toward the detector. At the receiving end, the high-power infrared sensor detects the infrared pulses. These pulses are amplified and filtered to reduce noise pickup, and they are then fed to a comparator after further amplification to restore thc pulses. These amplified pulses are then demodulated and fed to the up–down counter. The counting-up and counting-down take place when the moving unit travels forwards and backwards respectively. The outputs of the counters are input to the two eight-bit-latch stages. These are connected to the input port of the computer in a way which is similar to that which was used with the ultrasonic-feedback circuit to ensure the access of sixteen-bit output words to the input port of the computer.

A more sophisticated optical sensor which can be used for this purpose is made by Heidenhain[39] of Germany. It has a detector head which scans over a film. The output device is connected to the head by a long cable. The maximum traversing speed that it can measure is 1 $m s^{-1}$, and the permissible acceleration is 8 $m s^{-2}$. The weight of the scanning head is only 20 g.

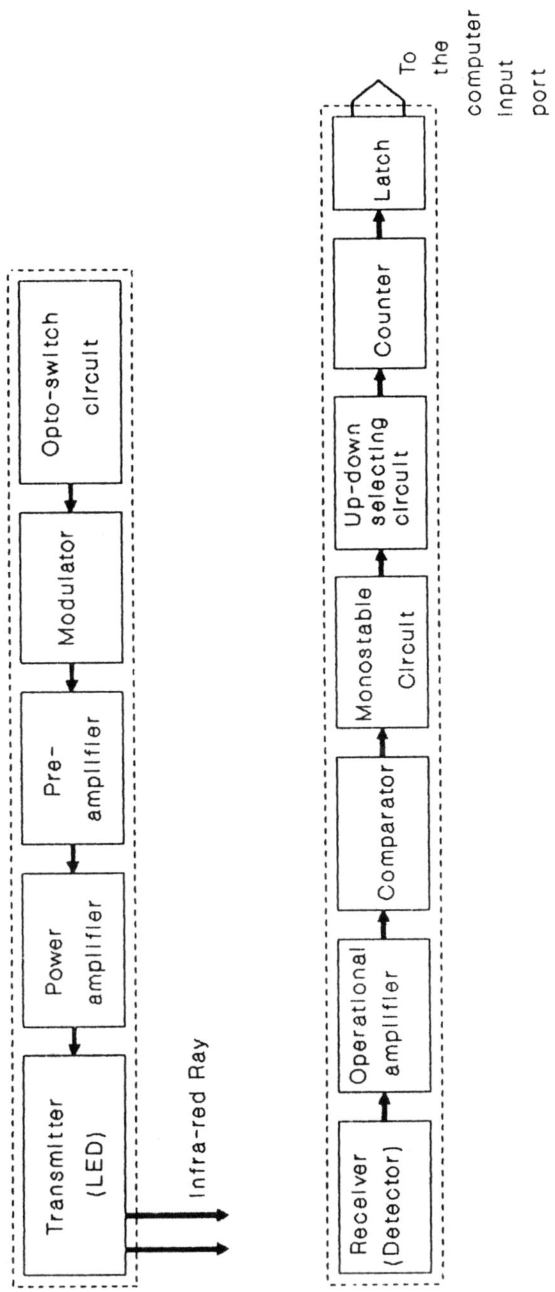

Fig. 6.6. A block diagram of the optical feedback circuit.

6.1.3 *Linear variable differential transformers*

Linear variable differential transformers (LVDTs) act as electromagnetic position transducers which produce an output signal which is proportional to the displacement of a moving magnetic core. Figure 6.7 shows an electrical circuit of a basic LVDT. It comprises a primary winding and two differentially connected secondary windings which are arranged concentrically about a central magnetic core.

The primary winding is normally excited from a fixed-voltage-and-frequency source. A symmetrically placed moving core induces equal voltages in the secondary windings, and the net output voltage is zero. When the core is displaced the flux linked by each secondary winding changes, resulting in an output voltage which varies linearly with respect to the core displacement.

Commercial LVDTs are available with a wide range of performance specifications, with the stroke length ranging from 0.5 to 500 mm, typical stroke/body-length ratios being of the order of 5:1, and the supply frequency ranges from 50 Hz to 10 kHz and above.

Now we will discuss the criteria[61] which are commonly used to assess the performance of LVDTs as position transducers.

(1) *Linearity* is defined as the maximum deviation of the output from a line of best fit within a specified displacement range. Typically, LVDTs have linearity errors of less than 0.5 per cent of their maximum output.

(2) *Sensitivity* is the output voltage for a given core displacement and a given input voltage. Typically this can be of the order of 100 mV per mm core displacement per volt excitation.

(3) *Resolution* is the smallest change in core position that can be detected. Theoretically this is infinitesimal; but in practice it is limited by the detector. Typically a resolution of 10 micrometres is achievable.

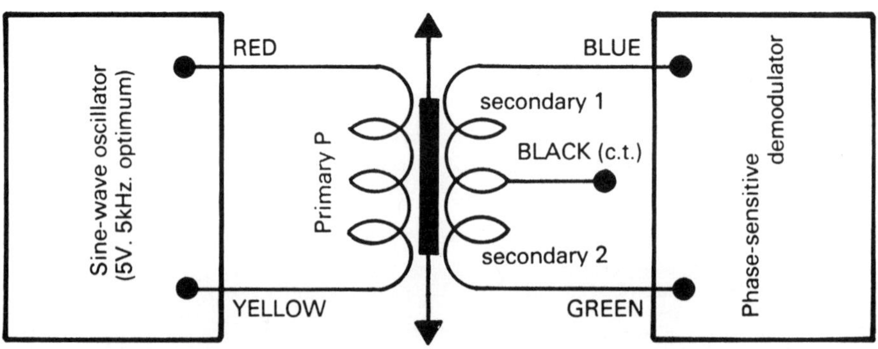

Fig. 6.7. An equivalent circuit of an LVDT.

(4) *Repeatability* is the consistency of the output. In theory, LVDTs are absolute-position transducers which offer infinite repeatability. However, in practice, the repeatability of the system is governed by the associated supply-and-detection electronics. Typically, the repeatability is of the order of millivolts, which corresponds to tens of micrometres, although this value is dependent on the sensitivity of the device.

Figure 6.8 shows a block diagram of an LVDT which was used to measure the position of the double-armature motor described in section 2.3.3.

6.1.4 *Inductively coupled sensors*

Inductively coupled sensors are noncontacting, inductively coupled, precision, scale elements that are used for extremely accurate measurements of displacements. They consist of a scale and a slider that are mounted with a small air gap between them. Inductive coupling between the Inductosyn conductor patterns bonded to the flat faces of the scale and the slider provide electrical signals that vary as a function of their relative displacement.[42] Electrical connections are made to one end of the scale, and also to the slider via a flexible cable. The slider can be directly mounted on the moving unit of the linear motor. The length of the scale equals the desired travel distance plus 114 mm.

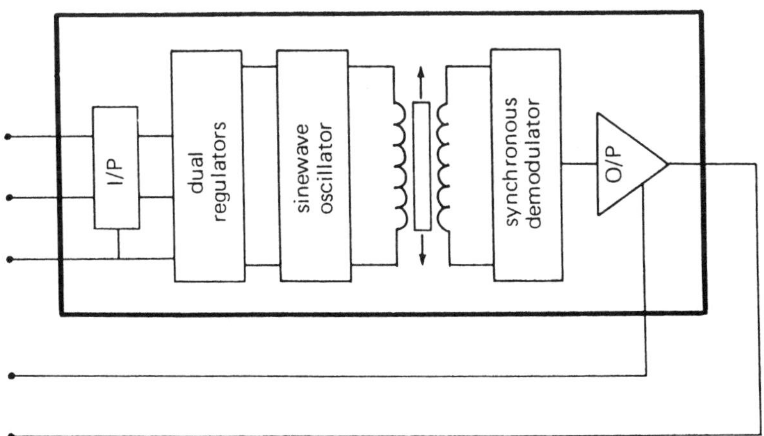

Fig. 6.8 A block diagram of a DCT 4000 LVDT.

6.2 The control system

A system used for controlling a 3 m long motor with 64 armature winding sections is discussed in this section.[3] This system controls the supply of electrical power to the 64 lines which are required to switch each individual section of the armature winding separately. It consists of a microcomputer, a control circuit, a switching circuit, and a sensor. Sixty-four light-emitting diodes are connected to the bases of the switching power transistors to indicate the position of the moving unit. Figure 6.9 shows a block diagram of the control system including some of the sensors discussed in Section 6.1.

In order to provide a smooth run for the carriage, two adjacent sections of the armature winding have to be simultaneously energized, over a certain period of time, especially at low speeds.

The control circuit consists of six 1-of-16 decoder/demultiplexers (type 74LS154N), 16 quad NAND gates (type 74LS00), two hex inverter-buffer/drivers (type 74LS06), and 64 npn transistors (type BC-108). Figures 6.10–6.12 show the circuit diagrams of the control circuit. A double-sided photoresistant copper-clad printed board can be designed and fabricated for mounting the elements of the control circuit.

The 74LS154N which is a 4-line to 16-line decoder/demultiplexer is designed to accept four-bit binary data and to provide 1-of-16 mutually exclusive, active, 'low' outputs. The enable gate has two ANDed inputs which must be 'low' to enable the outputs.

The 74LS154N can also be used as a 1-of-16 demultiplexer by using one of the enable inputs as the multiplexed data input. When the other enable is 'low', the addressed output follows the state of the applied data.

The quad NAND gates, which are shown in Figs. 6.10 and 6.11, provide the overlap energization of two adjacent sections of the armature winding. Half of these gates are used as NAND gates, while the other half work as inverters.

The hex inverter buffer inverts one of the enable inputs of each of the decoders shown in Fig. 6.10, and it also inverts the common line for the second enable input,which is shown in Fig. 6.11.

The outputs of the NAND gates shown in the control-circuit diagram in Figs. 6.10 and 6.11 are not large enough to saturate the high-current power transistors which are used as switches. Therefore, intermediate build-up current stages (see Fig. 6.12), which raise the base current of the power transistors in the switching circuit to a value high enough to ensure saturation, have been used.

Figure 6.13 shows the switching circuit which delivers electrical power to the individual sections of the armature winding. This circuit uses an MJ11015-type Darlington power transistor. It carries a maximum collector current of 30 A, and it dissipates 200 W.

In order to saturate the transistor, a large current is applied at the base,

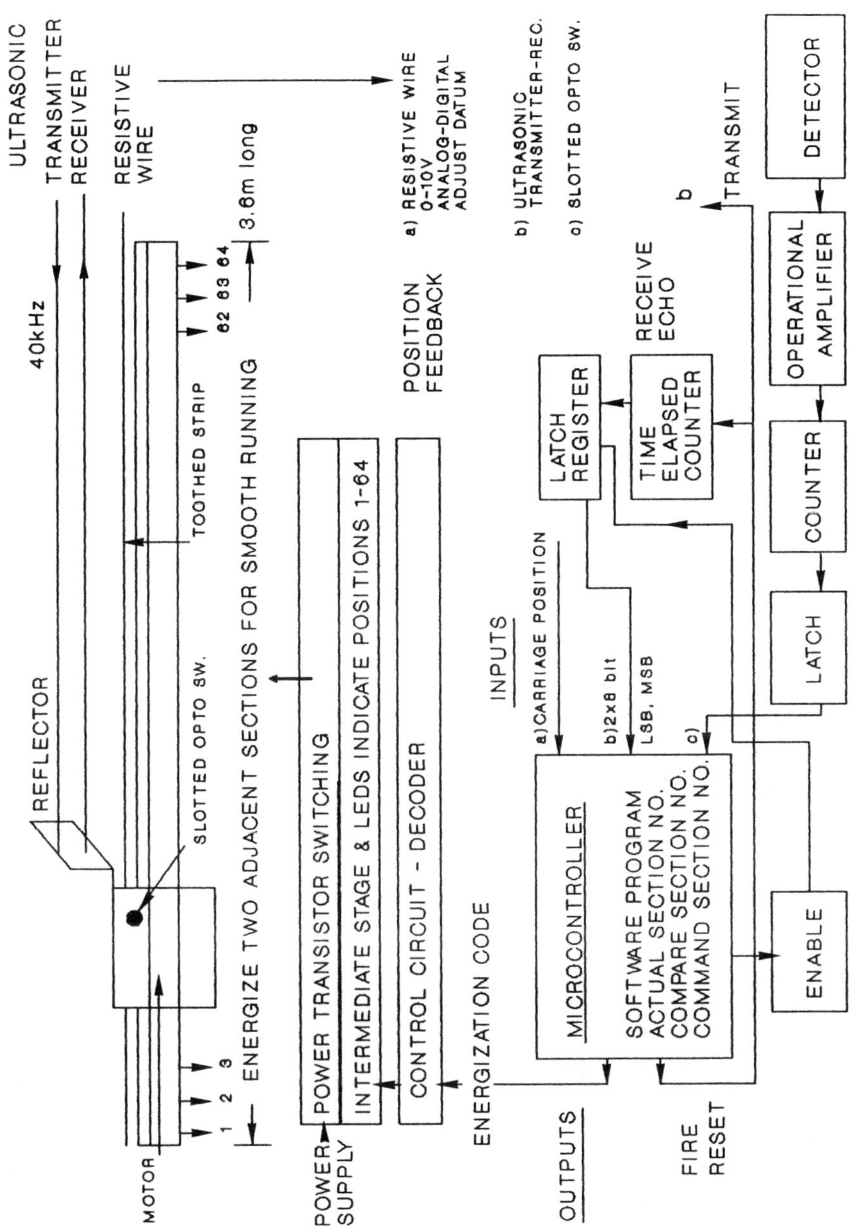

Fig. 6.9. A detailed block diagram of the control system, showing several types of position sensor.

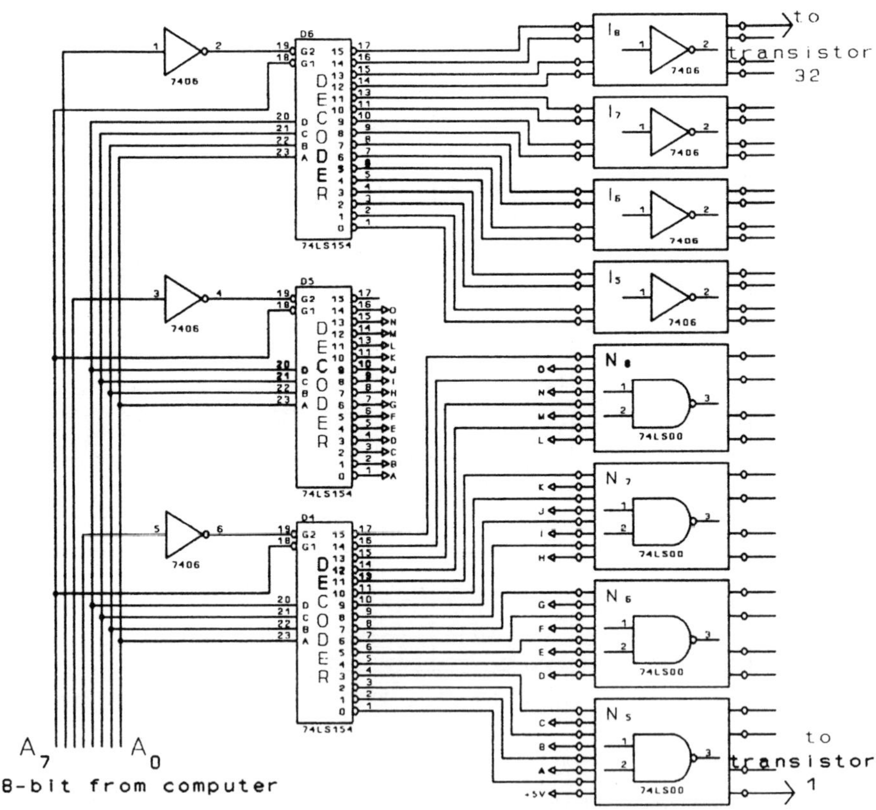

Fig. 6.10. The control circuit (continued in Fig. 6.11).

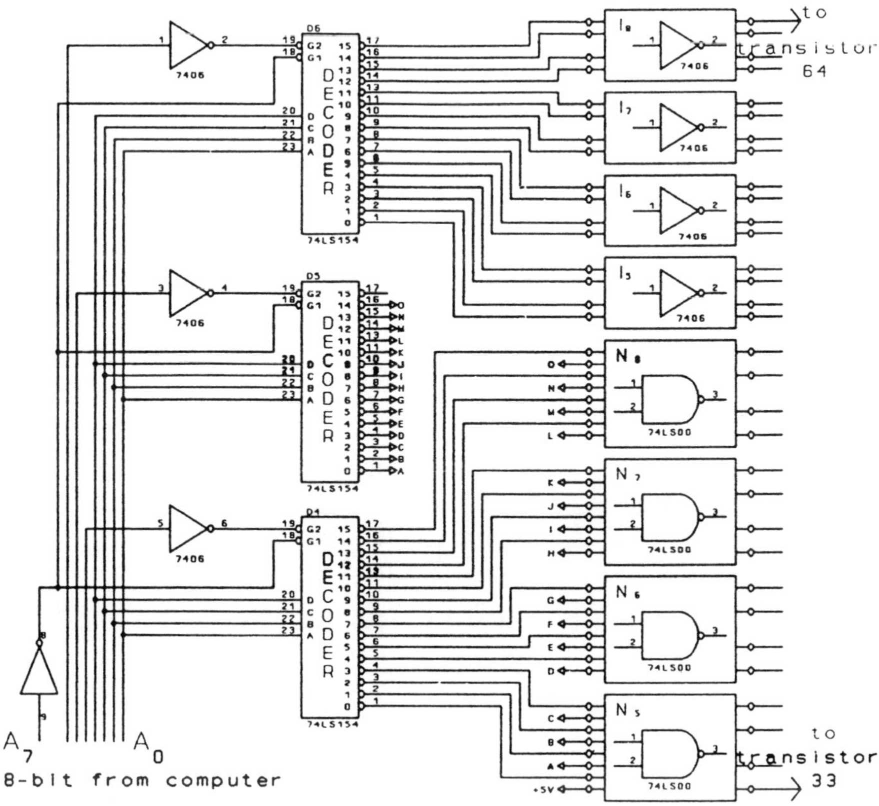

Fig. 6.11. The continuation of the control circuit.

turning the armature-winding sections on and off in sequence. The armature-winding sections of the motor are inductive, and they appear to be combinations of inductance and resistance in series. When the power transistor is connected to an armature-winding section, the inductance of the armature-winding causes a high reverse voltage to be induced in the winding when the transistor is switched off. This reverse voltage *Ldi/dt* may cause damage to the transistor if it is not suppressed. One method of suppressing this spike voltage, and of protecting the transistor, is to connect a resistor in series with a diode, as shown in Fig. 6.13. This damps the circulating current quickly. The voltage V_{CE} applied to the collector at turn-off in this configuration is

$$V_{CE} = E + IR_s + V_{DF} \ ,$$

where E is the supply voltage,
I is the current just before turning off,
R_s is the resistance of the suppressor resistor,
and V_{DF} is the diode's forward potential.

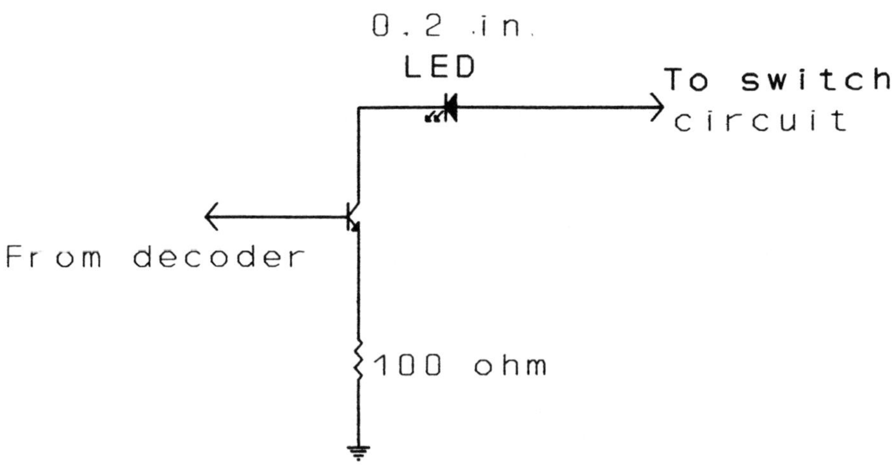

Fig. 6.12. One of 64 intermediate stages.

The diode and the resistor are in parallel with the armature winding and they protect the power transistor from the surge current during switch-off.

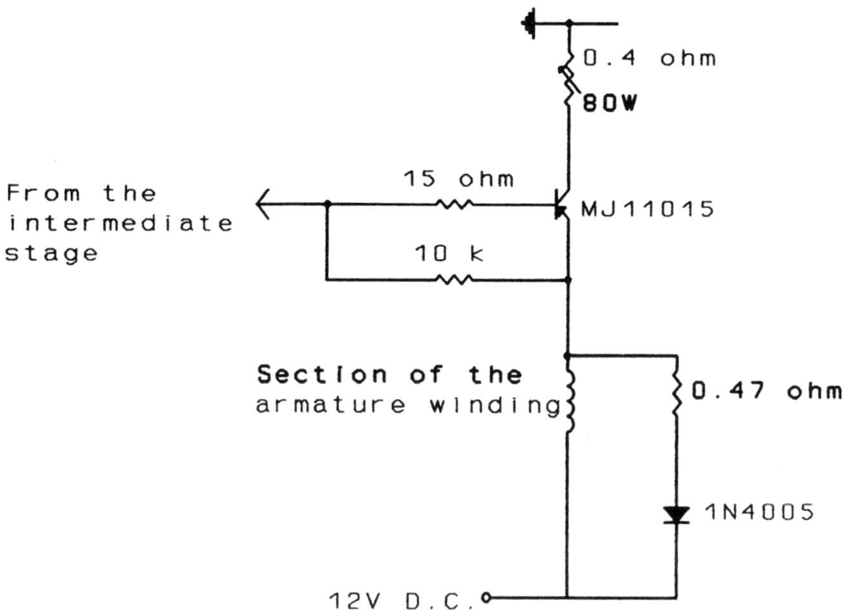

Fig. 6.13 One of 64 switching circuits.

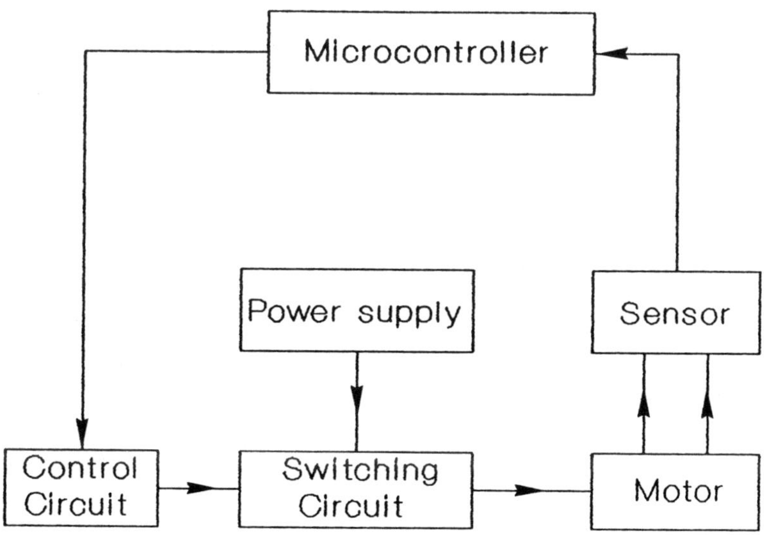

Fig. 6.14 A block diagram of the control system.

A block diagram of the control system is shown in Fig. 6.14. If we assume that the moving unit is at the start of the track, and that the microcomputer is programmed to move it to any other position, the output of the microcomputer has to be decoded by the 1-of-16 decoder/demultiplexer. The first output of the decoder becomes ‘low’ while the other outputs become ‘high’, and when these signals are fed to the inverters, only one inverter will have a ‘high’ output. This saturates the intermediate transistor, which then drives the switching power transistor into saturation. Hence, the collector current of the switching transistor flows via the first section of the armature winding. The other sections remain unenergized due to the cut-off state of the intermediate and switching transistors.

The sensor detects the signal voltage which represents the position of the moving unit, and the commands of the microcomputer remain the same, so that the moving unit keeps on travelling in the same direction until the rear pole completely faces the second section of the armature winding. The command voltage from the microcomputer, at this point in time, changes so that the second output of the decoder becomes ‘low’. This signal eventually energizes the second section of the armature winding after passing through the inverter and driving sequential transistors into conduction. The moving unit travels section by section, until a stop command is achieved when it reaches the pre-programmed position.

In order to reverse the moving unit, the section of armature winding which is under the front poles should be energized. The reverse motion can be used to brake the carriage when it travels in the forward direction.

Sections 1–32 of the armature winding are energized by controlling the address and enable data inputs which are fed to the control circuit of Fig. 6.10. At any particular time, both of the enable inputs G_1 and G_2 are ‘low’ for at least one of the decoders D_1, D_2, and D_3. In this case, the most significant bit of the microcomputer output A_7 should be set to ‘low’. Hence the enable input G_1 to the three decoders D_4, D_5, and D_6, shown in Fig. 6. 11, becomes ‘high’. This sets the outputs of all the decoders of this circuit to ‘high’.

To energize sections 33–64 of the armature winding, A_7 is set ‘high’, then the decoders D_1, D_2, and D_3 are enabled, while at least one of the three decoders D_4, D_5, and D_6 controls the energization of the armature-winding sections.

Jerky motion occurs when the carriage moves from one section of the armature winding to the other. In order to provide a smooth run of the carriage, the following section should be energized *before* the front magnet pole of the carriage has completely passed onto the following section. This means that there is an overlap time during which two adjacent sections of the armature winding are energized simultaneously.

The six decoders have common address inputs, so the addressed outputs of the decoder can be controlled by changing the enable inputs. A 'low' A_7 and a 'high' A_4, A_5, or A_6 disable D_1, D_2, or D_3, respectively, as can be seen in Fig. 6.10, while a 'high' A_7 and a 'high' A_4, A_5, or A_6 disable D_4, D_5, or D_6, respectively, as can be seen in Fig. 6.11.

Table 6.1 shows the eight-bit parallel inputs, and the numbers of the corresponding energized sections of the armature winding. For the second and third sections of the armature winding to be simultaneously energized, A_5 and A_6 have to be held 'high', while A_4 and A_7 are held 'low'. The enabled inputs A_4, A_5, A_6 and A_7 enable D_1, D_4, D_5, and D_6 and they disable D_2 and D_3.

The NAND gate N_1 receives two inputs, $\overline{1}$ from D_1, which is 'high', and $\overline{0}$ from D_2, which is 'low'. The output of N_1 is therefore, 'high'. This drives the intermediate transistor T_3 into saturation. Thus T_3 drives the power transistor into conduction, and the third section of the armature winding is eventually

Table. 6.1 Energization codes for the first six armature sections

A_7	A_6	A_5	A_4	A_3	A_2	A_1	A_0	Energized section
L	L	L	H	L	L	L	H	1
L	L	H	H	L	L	L	H	1,2
L	L	H	L	L	L	L	L	2
L	H	H	L	L	L	L	L	2,3
L	L	L	H	L	L	L	H	3
L	L	H	H	L	L	L	H	3,4
L	L	H	L	L	L	L	H	4
L	H	H	L	L	L	L	H	4,5
L	L	L	H	L	L	H	L	5
L	L	H	H	L	L	H	L	5,6
L	L	H	L	L	L	H	L	6
L	H	H	L	L	L	H	L	6,7

energized. At the same time $\overline{0}$ from D_3 is held 'low' and, after being inverted, the transistor T_2 is driven into saturation, which switches the power transistor on and energizes the second section of the armature winding.

6.3 Speed control

The speed of a motor can be controlled by regulating the armature current or by changing the overlap time when two adjacent sections of the armature winding are simultaneously energized. In the former case, a quad, single-pole, single-throw, dielectrically isolated analog switch, with integral data latches and a group of resistors, can be connected to the current-programming and negative terminals of a power supply, as shown in Fig. 6.15. This power supply has auxiliary terminals for remote control of the output current, and a variable resistor with a resistance from 0 to 4.7 kΩ produces an output current of zero to full scale. Software controls the analog switch, which, in turn, connects resistors with different resistances to the power supply.

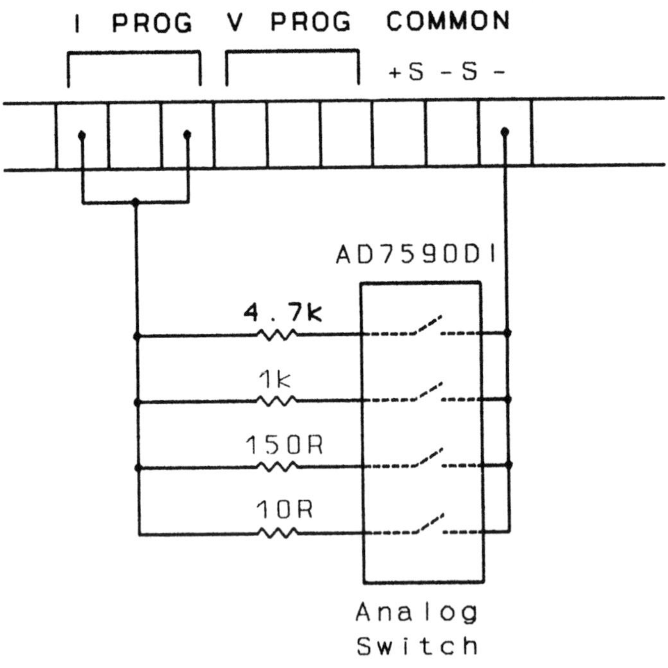

Fig. 6.15. The remote control circuit of the power supply.

For the second case, examine Fig. 6.16. The top waveform shows the energization period $T_{ON1} - T_{OFF1}$ for the first section. The speed of the motor is regulated by shifting the switching-on time, T_{ON2}, for the second section. The period of overlap, T, is normally less than the energization time for each section, otherwise a reversal force acts as a brake on the moving unit, thus decreasing its speed.

6.4 Position control

The microcomputer normally employs a self-contained program called an interpreter which analyses each line of instruction entered by the user and then performs the desired operation. The Basic language, which is used for writing the program, is translated into machine language.

For high-speed control and measurements an assembly language is used while the Basic language is used to perform less time-dependent tasks.

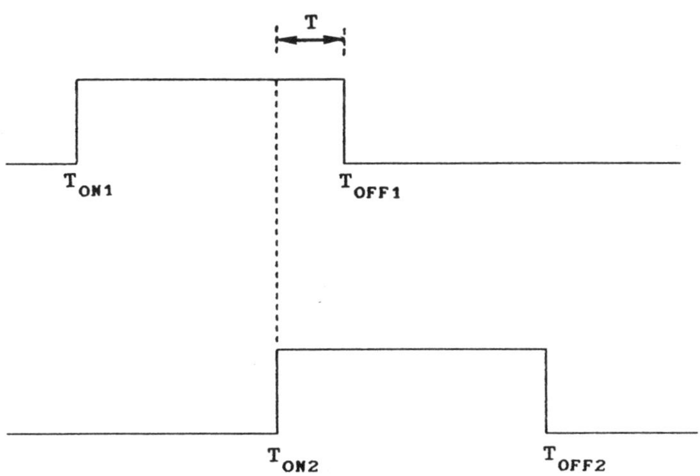

Fig. 6.16. Timing waveforms for energization of two adjacent sections of the armature winding.

6.5 Control software

The logic for the control program is shown in the flow chart in Fig. 6.17. The

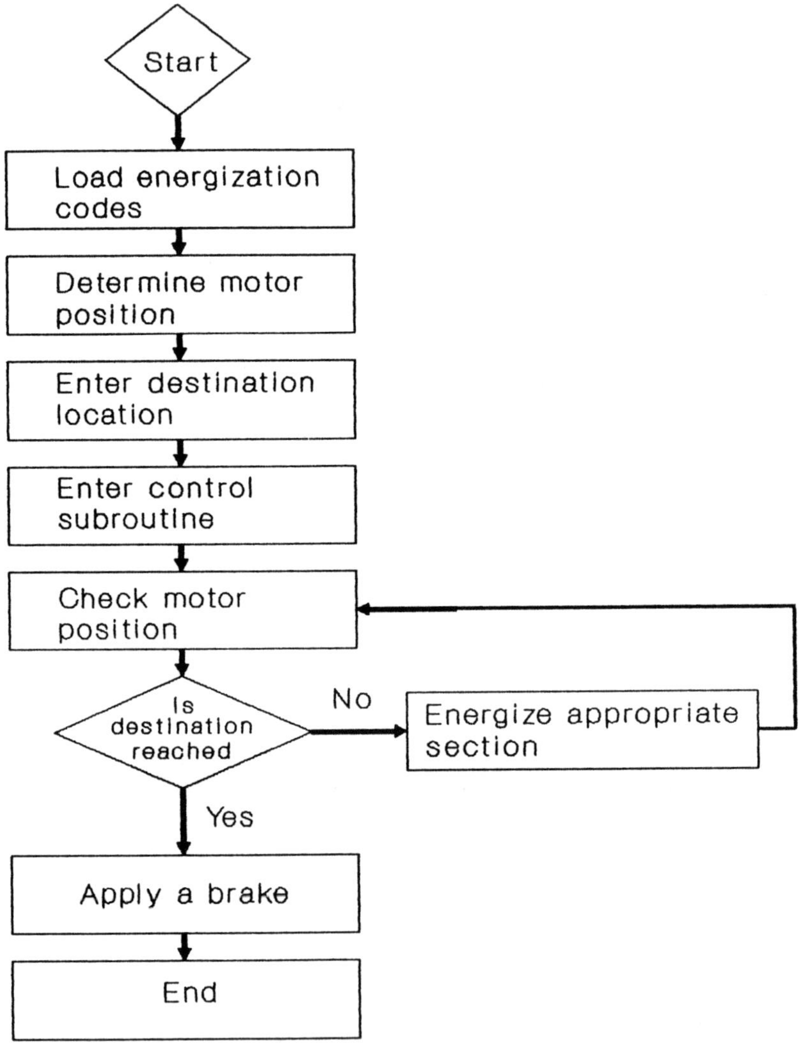

Fig. 6.17. A flow chart for the control program.

main program is used to load memory locations with eight-bit energization codes, and the respective representation values of track position by using the 'poke' statement. The program also loads the memory locations with the relevant codes in order to stop the moving unit of the motor. The main program then enters a subroutine which checks the motor position and controls the energization of the armature sections until the destination is reached. When the moving unit arrives at its preprogrammed destination, a brake is applied and the subroutine returns control to the main program.

Improvement to the control software is possible by simply increasing the complexity of the closed-loop software. As the position of the motor has to be monitored at a high sampling frequency, the use of a high-level language, such as Basic, is unacceptable due to the speed limitations.

7

DESIGN AND ANALYSIS

An engineer designs a motor according to the specifications provided by the customers. For linear motors, the main specifications are the sensitivity of the motor (thrust coefficient), the output force, the output power, the efficiency, the velocity-versus-time characteristic, and the power-to-weight ratio. When determining the power-to-weight ratio, it is normal practice to take into account the weight of the whole moving unit, and also the weight of the part of the stationary unit which is enclosed by the moving unit.

In the first two sections of Chapter 4, expressions for the static thrust, eqns (4.12), (4.13), (4.18), (4.19); velocity, eqns (4.27), (4.65), (4.75); developed force, eqns (4.50), (4.60); output power, eqns (4.50), (4.58), (4.59); and efficiency, eqns (4.27), (4.65), (4.74) were derived for motors with different configurations. In the Appendices these parameters are given for several square-armature d.c. linear motors. The parameters, in general, for linear motors are comparatively poor due to inherent design requirements such as wide air gaps, but then they are low-cost, space-saving, and nearly maintenance-free motors.

There are several possible analyses of any device which can help a designer to improve the performance of the device. In this chapter more emphasis has been given to measurements and computations than to the design. We analyse the experimental and computed results of the static thrust and the flux density distribution of several motors. Experimental results for dynamic performances are also presented and analysed. Since d.c. linear motors with permanent magnets are novel types of motor, some of the experimental methods and test rigs are new, and they need to be described here. All the formulae needed for designing a d.c. linear motor are presented in Chapter 4.

7.1 Design of a multiwinding motor

Firstly, let us consider a brushless d.c. linear motor consisting of a multiwound square armature and a four-pole moving unit. This is a machine with a long traverse, and the armature is chosen to be long and stationary; the short moving unit is the source of the magnetic flux. The number of poles can be varied by even numbers, and for a specific power requirement a larger number of magnets with smaller volumes will produce the same power, and since the price of rare-earth permanent magnets increases exponentially with the volume, a greater number of poles is preferable for low-cost machines.

From a knowledge of the flux density distribution one can optimize the shape and size of the various magnetic materials which provide most of the magnetic path of the device. The cross-sectional area of the armature core is an important factor in the design, since the static thrust is directly proportional to the square of this area. Again, the conductor size depends on the power and on the capability of the control system to drive the armature current. The sections of the armature winding are energized, in turn, for an insignificant time and the current can therefore be chosen to be several times the continuously rated current of the armature conductor. For this reason no heat sink is required, and sections with several layers of winding are possible.

The length of the air gap between the magnet poles and the armature should be uniformly maintained on either side of the armature, since unequal air-gap lengths produce an unbalanced force of attraction between the magnet poles and the armature on either side. This results in a force component which is normal to the track length, and this affects the motion of the carriage.

7.1.1 *Armatures*

The length of the armature determines the travelling distance of a motor. In this particular case the armature core is made by joining two sections of 1.8 m long mild-steel square bars with a 25 mm^2 cross-sectional area. This area mainly depends on two factors: the height of the permanent magnets, and the fact that the core should not be saturated with flux.

A single layer of 14 standard wire gauge (s.w.g.) enamelled copper wire is closely wound on the armature core. A thin layer of glass-cloth tape (130 °C rated) is wrapped around the full length of the core to provide extra insulation between the armature windings and the core before the wire is wound on to the bar. The diameter of the wire is chosen to give the required number of turns per metre length of the armature, which is proportional to the developed thrust. The smaller the diameter of the wire the greater is the number of turns per metre length for a given length of armature-winding section, and hence the greater the force the less is the current-carrying capability. By using 14 s.w.g. enamelled copper wire, a compromise is achieved between the force produced by the motor and the maximum current which can pass through the windings without causing any damage. The ends of the copper wire are mechanically secured to the track base and they are electrically connected to the power supply through the switching circuit which was discussed in the previous chapter. This forms one section of the armature winding, and there are 64 such sections on the core in this motor. The armature is mounted on a wooden plinth supported by Tufnol blocks at 1 m intervals.

7.1.2 *Field systems*

The magnetic-field source is housed in the moving part of the d.c. linear motor. The whole system is held by a shell made from a cylinder of Polypenco material with a slit in the lower part through which the wire ends of the armature-winding sections pass.

High performance neodymium–iron–boron magnets save space, and, mainly for this reason, they have been used to generate the magnetic field. Specifically, Crumax-301 ($Nd_{15}Fe_{77}B_8$) magnets are chosen for this purpose. The width and depth of each of the four magnets are 26 and 25 mm respectively. The amount of radial flux which interacts with the current-carrying conductors is directly proportional to the area of the magnets and hence the area is a factor in determining the force developed. Yokes made from steel bars with a cross-sectional area of $12 \times 25\,\text{mm}^2$, known as back irons, are bonded to the back of the magnets on either side of the armature, and this completes the magnetic circuit.

In the design of this motor, the relevant equations from Chapter 4 were used. The motor is designed for a specified thrust and for the largest possible output-power-to-weight ratio.

7.2 Design of a double-armature motor

The motor described in Section 2.3.3 has a rectangular double-armature core made of two long and two short mild-steel bars joined together. There are gaps or inclusions of nonmagnetic material along the bars to avoid magnetic saturation, and to achieve a better electrical time constant. This motor can use two different types of winding: (a) a continuous winding forming two separate coils, C_1 and C_2, that can be fed with a current with the same magnitude, I_a, but with different polarities, and (b) several sections of winding each with the same number of turns, each being fed with switched currents which characterize the behaviour of a linear stepping motor. Configuration (a) has been adopted for this particular motor. Its slider is rectangular and it is made of mild steel. It is provided with two linear bearings for supports, and it is guided by two cylindrical rods.

In order to generate the maximum magnetic flux with only a slight increase in the volume of the motor, seven neodymium–iron–boron magnets are used. They are held by the slider, and they surround the armature bars, as shown in Fig. 2.22. Using such an arrangement, the active length of the armature conductors is increased. As a result, the motor can develop a larger thrust for a lower current. In other words, a larger sensitivity of the motor (thrust coefficient), which is the ratio of the thrust to the current, can be achieved. This feature also requires a smaller power supply, and it reduces the power dissipation in the winding. The power reduction is particularly important in this case, since electrical-power losses raise the operating temperature of motors, and might cause irreversible changes in the magnets. The magnets have a remanent flux density of $B_r = 1.15$ T, a coercive force of $H_c = 860$ kA m^{-1}, and the maximum energy product is $(BH)_{max} = 246$ kJ m^{-3}. These magnets have a nickel-based coating to avoid any corrosion. The auxiliary magnets are placed at the inner side of the slider. The main magnet, which is placed between C_1 and C_2, is mechanically fixed to the slider by two pieces of nonmagnetic material. The magnetic lengths of the magnets are such that they compensate for the demagnetizing effect of the air gaps along the individual magnet paths.

In Section 4.1 the expressions for force, velocity, and power for this motor are given in eqns (4.19), (4.20), and (4.21) respectively. If we adopt a linear magnetic-circuit model for the motor, it is possible to establish that the slider 'sees' two armature reluctances in parallel through the air gap. One of the reluctances, $\Re_a$, is due to the magnetic path in one-half of the motor and the other, $\Re_b$, is due to the other half. Both depend on the position of the slider. The value of $\Re_a$ in parallel with $\Re_b$ is given by

$$\Re_{ab} = \left[\left(\frac{r+2x}{\mu\, l_1^2} + \frac{2l_0}{\mu_0 A_0} \right)^{-1} + \left(\frac{r+2l_t-2x}{\mu\, l_1^2} + \frac{2l_0}{\mu_0 A_0} \right)^{-1} \right]^{-1} . \quad (7.1)$$

In eqn (7.1), r is a constant equal to $3l_1 + 2l_m + 2l_g$, where l_m is the length of the side magnets, l_g is the air-gap length, μ and μ_0 are the magnetic permeability of the core and air, respectively, l_1 is the active conductor length, l_0 is the length of each of the nonmagnetic gaps along the armature core, A_0 is the cross-sectional area of the gaps, and l_t is the travel length. The distance between the centre of the slider and one end of the long bars of the stator is given by x.

From eqn (7.1), the magnet load line in this particular case is given by

$$\frac{B_m}{H_m} = -\frac{\mu_0 l_m C_f L_f}{l_g + 2\mu_0 A_0 \Re_{ab}} , \quad (7.2)$$

where C_f and L_f are the fringing-flux and the leakage-flux factors, respectively, for the magnetic flux provided by the magnets; C_f and L_f depend on the geometry of the motor. They can be estimated according to values given in the literature or they can be obtained with the aid of finite-element analysis. It can be proved from eqns (7.1) and (7.2) that $\mathfrak{R}_{ab}$ has a parabolic behaviour with respect to x. Consequently, the magnetic flux provided by the magnets is a minimum at the centre of the travel length and a maximum at the extremes of the armature core, as is the static thrust. This nonlinearity can be minimized using either a material with a larger permeability or with a nearly constant reluctance. The second option is feasible for short-stroke motors in which the reluctance of the air gaps is much larger than the reluctance of the ferromagnetic materials used for the armature. The design involves the two equations above, conditions like maximum core saturation, travel length, maximum desired static thrust, and the use of an interactive calculation process. It is thus possible to obtain the dimensions of the magnets, the armature cores, and the air gap, and also the number of turns per armature winding section. For a particular thrust, each section has 1140 turns distributed in four layers, the air gap is 4 mm long, the nominal armature current is 1.5 A, each auxiliary magnet is 5.2 mm thick, and the main magnet is 10.4 mm thick. At the middle of the travel length according to the model, the theoretical flux density in the air gaps between the magnets and the sides of the armature core they face is $B_g = 0.51$ T.

7.3 Measurement of the thrust, speed, and field

One of the most important parameters of a d.c. linear motor is the static thrust. It can be measured by using the system shown in Fig. 7.1. A load cell is fixed at one end of the track, and its stem is attached to the moving part of the motor, as is shown in Fig.7.1. The armature winding is energized for a short period of time. A voltage proportional to the force developed by the motor is detected at the output of the load cell, and it is fed to an amplifier before recording. The static thrust may be obtained for a range of armature currents and air gaps between the armature and the poles.

7.3.1 *An apparatus for thrust measurement*

A specially designed load cell may be used for measurements of static and dynamic loads. In general, this cell consists of four active and four temperature -compensating gauges (electrical-resistance strain gauges) connected in a four-arm Wheatstone-bridge configuration. By applying a force to the stem of the load cell, an out-of-balance error signal is produced due to the change of the resistance in the strain gauges. The sensitivity of such a load cell, which is capable of measuring a force of up to 20 kg, is typically 2 mV per excitation volt, the accuracy is 0.1 per cent and the hysteresis of the output is within

0.05 per cent. Figure 7.2 shows a test rig for measuring the thrust in a double-armature d.c. linear motor with the aid of such a load cell. The slider of the motor is mechanically connected to the load cell. The recommended excitation is 10 V d.c., and the input and output impedances are 415±15 Ω and 350±3 Ω respectively.

7.3.2 *Speed measurement*

The speed of d.c. linear motors is another important parameter from which the dynamic characteristics are obtained. It may be measured using various techniques, some of which are very accurate, and others are not. The method used here is fairly accurate; it uses a microcomputer in conjunction with an optical-feedback system to record the speed. A high-level language is used to load memory locations with eight-bit energization codes and with the respective values of the track position. Values representing different fixed points on the track are loaded into several memory locations. A subroutine continuously monitors the position of the motor, and it outputs eight-bit winding codes according to the position of the motor. This subroutine also acts as a counter.

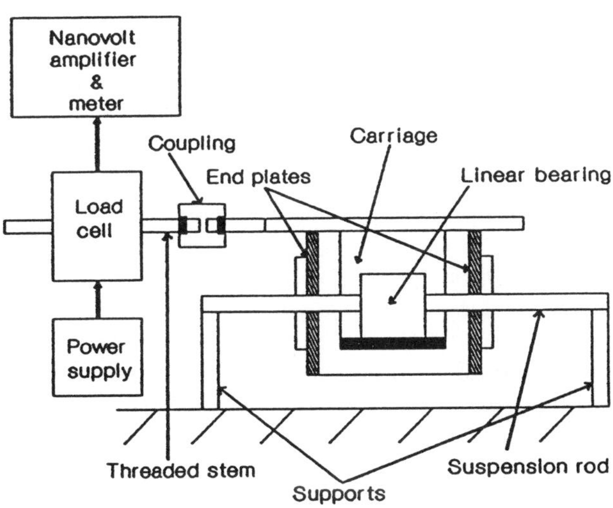

Fig. 7.1. Measurement of the static thrust of a single-armature motor.

Fig. 7.2. Measurement of the static thrust of a double-armature motor.

A register pair is used as a counter to determine the travel time between six fixed points on the track, so that the moving unit of the motor, when it is at a certain position, initiates the counter by using 'compare' and 'jump' statements. The counting continues until the moving unit reaches the next point. The count cycle then stops, the loop cycle begins, and the data is loaded into memory locations for future use. Another count cycle then starts, and the procedure is repeated. Once the moving unit reaches the destination, the subroutine returns control to the main program. The overall time taken for the execution of each count cycle is calculated. A 'multiplication' statement in the main program is used to compute the travel time of the moving unit of the motor between two adjacent points, and, since the distances between these points are known, the speed of the moving unit can be calculated.

7.3.3 *An apparatus for flux measurement*

Analysis of the flux is also important in determining the performance of any electrical machine. The flux distribution in a d.c. linear motor can be determined by several different methods of magnetic measurement. Instruments such as Grassot fluxmeters, Hall-effect Gaussmeters, and ballistic galvanometers are commonly used for such measurements in conjunction with *B*-coils, *H*-coils, and Chattock potentiometers.

7.3.4 *Measurement of the axial flux in an armature*

The magnetic circuit of a d.c. linear motor usually consists of a field source, the air gap, the back iron, and the armature core. The field source and the armature winding have different contributions to the flux distribution in the motor which depend on factors such as the armature current and the air-gap length. The flux at any part of the motor is the vector sum of the flux produced by the field source and the flux produced by the armature current.

In order to measure the flow of axial flux inside the armature core which is produced by the field source, no current is supplied to the armature winding (so that thc moving unit does not move). Several search coils, made of thin, enamelled, copper wire, are closely wound around the armature winding. Since both the fluxmeter and the galvanometer respond to changes in the flux linkage through the search coil, the field source can be suddenly removed to a point where it has no linkage with the search coils.[13] The axial flux generated by the armature excitations is easily calculated. The vector sum of the flux produced by the field source, and the flux produced by the armature excitation, gives the value of the total axial flux. This procedure is repeated at different positions of the motor. Thus, a profile of the axial flux flow can be obtained for a particular motor. The experiment is normally repeated for various air-gap lengths.

7.3.5 *Measurement of the air-gap flux due to a field source*

In the air gap, the radial flux due to the field source can be measured in several ways. An array of Hall-effect elements is used to determine the radial-flux variation on the surface of the armature winding. When a bias current flows, a signal voltage is generated at the output terminals of each of the elements which are in a magnetic field. This voltage is proportional to the product of the magnetic field strength and the current, and it is perpendicular to both of them.

Instead of using an array of Hall-effect elements for flux-variation measurement, only one search coil can be used to obtain the average radial flux. The coil is again made of thin, enamelled, copper wire and it has one turn. Its shape is rectangular, and it covers the same area as a magnet pole. The search coil is placed in front of each of the poles in turn, and the field source is suddenly removed to a place which has a negligible field strength compared with the initial position. This produces a change in the flux linkage. A galvanometer can again be employed to obtain the value of flux.

The use of search coils to measure the radial flux can be avoided by taking direct measurements with the aid of a Gaussmeter. In this case the readings can be taken anywhere along the length of the air gap, from the surface of the armature winding to the magnet pole surface, with the aid of a rectangular detector made from a 6 mm long, 3 mm wide, and 0.8 mm thick piece of indium arsenide mounted on a 76 mm long probe. Thus a very good knowledge of radial flux distribution in the airgap can be obtained.

7.3.6 *Measurement of the air-gap flux due to an armature current*

A pancake-type search coil with 300 turns, a 1 cm overall diameter, and a 0.5 mm thickness made from 50 s.w.g. enamelled copper wire is placed on the armature winding in such a way that the area of the search coil is parallel to the armature axis, and consequently it is perpendicular to the direction of the flux flow in the air gap. The moving unit is kept in a fixed position, and the armature current is suddenly reversed to produce a deflection on a galvanometer. The deflection is proportional only to the magnetic flux caused by the armature current, since there is no change in flux due to the field source linking the source coil, and consequently the field source has no effect on the galvanometer readings. This procedure can be repeated at different positions along the armature axis in order to measure the air-gap flux density due to the current in the armature winding along the whole length of the motor.

7.3.7 *Measurement of the total flux*

In order to obtain the total flux, a single-turn search coil made from 48 s.w.g., enamelled coppper wire and a demagnetizing coil of 120 turns (48 s.w.g.) are wound round the middle of one of the yokes in the moving unit of the motor. The demagnetizing coil, which is in series with an ammeter, is connected to a variable power supply through a double-pole, double-throw switch.[14] The secondary of a standard solenoid, which is in series with a ballistic galvanometer, is connected to the search coil, as is shown in Fig. 7.3.

A Hall-effect Gaussmeter with its thin sensor at the end of its probe is placed in the air gap between the armature winding and the magnet pole. In the absence of any current in the demagnetizing coil, the magnetic flux in the air gap between the armature winding and the magnet pole at points a and b, is measured as H_a and H_b respectively. The magnetic fields at these points can be changed by 10 per cent of the original values by varying the demagnetizing current I_d, so that the magnetic field at points a and b becomes

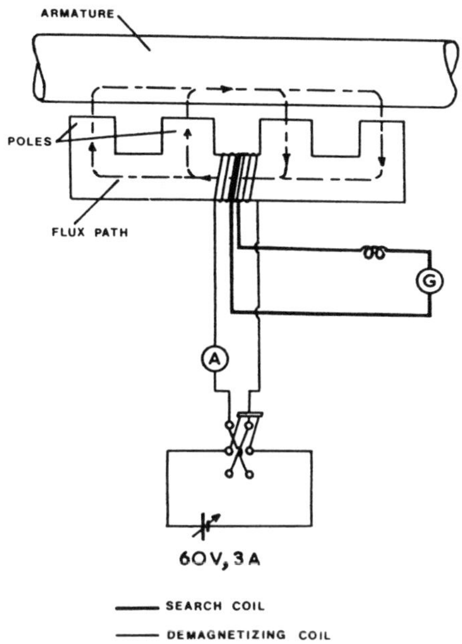

Fig. 7.3 Measurement of the total flux.

$H_a + \Delta H_a$ and $H_b + \Delta H_b$ respectively. By reversing the switch the demagnetizing coil can again be energized to I_d, but the direction is reversed. Magnetic fields with magnitudes of $H_a - \Delta H_a$ and $H_b - \Delta H_b$ are produced at points a and b respectively. Thus the reversal of the demagnetizing current I_d produces a magnetic flux change of $2\phi_d$ in the yoke. The flux ϕ_d flows out from the yoke through the pole to the armature and it returns to the yoke through the opposite pole. The total flux flow in the yoke[13] is given by

$$\phi_t = \phi_d \frac{H_a}{\Delta H_a} = \phi_d \frac{H_b}{\Delta H_b} \tag{7.3}$$

By using this procedure, the total flux flow in the yoke can be determined.

7.4 Computation of the field and force

In order to predict the distribution of magnetic flux in the d.c. linear motor, it is necessary to employ mathematical techniques which allow the flux density to be computed in three dimensions, or in two dimensions for symmetrical cases. It is possible to optimize the design by making changes in the motor geometry from a knowledge of the magnetic field distribution.

In recent years the finite-element method has been accepted in electrical engineering as an extremely powerful tool for the analysis of complex electromagnetic field problems. Ideally, any machine should be analysed using a three-dimensional software package, but two-dimensional programs are adequate for some geometries. Two-dimensional packages require less modelling, central-processing-unit time and memory space than three-dimensional packages.

7.4.1 Analysis of fields by a two-dimensional software package

Consider the motor shown in Fig. 2.20. In this case the permanent magnets and the armature have exactly the same depth, 25 mm. The depth of the magnets is the third dimension for the model. The reluctance for the leakage flux is very large and it keeps the flux distribution along the depth of the magnet uniform. A two-dimensional software package can, therefore, be used for the analysis of such a motor. Such a package consists, in general, of two separate parts, the preprocessor and postprocessor and the main analysis programs. The preprocessor and postprocessor is an interactive program which is used to define the geometry of the problem and to interrogate and display the solution. The static program solves the problems. All the programs are based on a differential-operator finite-element method which requires that the problem geometry is discretized into small elements, and that it is terminated by boundary conditions.

For magnetostatic problems with permanent magnets the equation

$$B = \mu(H - H_c) \tag{7.4}$$

is used to link the flux density B, the field strength H, and the coercive force H_c.

The area of a device over which the magnetic field distribution is to be computed is divided into regions with different material properties: conductors with defined current densities, irons with permeabilities and permanent magnets with appropriate B–H curves. In regions where the vector potential changes slowly, a coarse mesh is chosen, whilst a fine mesh (a dense cluster of small elements) is employed for those areas where severe changes take place. The permeability and the strength of the permanent magnet in nonlinear solutions are obtained from appropriate B–H curves which are provided by the software package. The B–H data for the permanent magnet must have the first point at $B = 0$, and the value of H for this point gives the coercive force, H_c, of the permanent-magnet material, while the first values of B and H for the steel yoke are both zero.

In order to reach a solution, about 20 iterations, each of which typically takes 90 minutes, are required.

7.4.2 *Computation of force by a two-dimensional software package*

The mechanical force developed by d.c. linear motors can be evaluated by analysing the magnetic field distributions of the devices. Several methods can be used, and all these methods should give the same result provided there is no computational error. However, because of the numerical approximations which are made in the individual solutions of the field equations, and in the distribution of the force itself, all these methods may not give identical answers.[15] Two methods which can be used for the evaluation of the mechanical force produced by motors are outlined below.

The virtual-work method. When a body is slightly displaced from its rest position, the force exerted on it may be evaluated by determining the work done to achieve this displacement. This work must equal the change in the energy stored in the motor, provided there is no friction; so if the stored energy is evaluated for the motor at two separate positions, then the force is equal to the difference between the two energies divided by the distance between the two positions, provided that the force does not change significantly during the motion; thus, this calculation is valid for small displacements only. Smith and Appel[74] have computed the thrust of a permanent-magnet linear actuator using this method.

The Maxwell-stress-tensor method. The net force can be found by calculating the local stresses at all the points on a boundary surface and then summing these stresses by a surface integral. The elementary force-density expression which relates the force-density vector, $\vec{f}$, to the current density, $\vec{J}$, and the flux density, $\vec{B}$, is given by

$$\vec{f} = \vec{J} \times \vec{B} \ . \tag{7.5}$$

In order to determine the force on a rigid body, the surface integral should enclose the body, and this surface is usually placed in the air region surrounding the object on which the force is to be calculated since air does not differ from free space in its magnetic properties, and it does not carry currents. From eqn (7.5), the net force on a conductor, viewed as a rigid body, is

$$\vec{F} = \int_{v} \vec{J} \times \vec{B} dV \ . \tag{7.6}$$

This integral is implemented as a built-in numerical routine in some finite-element packages.

In general, calculations of the Lorentz force employ this integral. However, to simplify the overall force calculation, this volume integral can be reduced to a corresponding surface integral by using the Maxwell stress tensor $\vec{T}$.

$$\vec{F} = \int_v \vec{J} \times \vec{B} dV = \frac{1}{\mu} \int_v div \vec{T} dV = \frac{1}{\mu} \int_s \vec{T} \cdot dS \; . \tag{7.7}$$

The calculations are performed in a two-dimensional coordinate system in which the surface integral becomes a contour integral and the tensor matrix reduces to the form

$$\vec{T} = \begin{bmatrix} \left(B_x^2 - 1/2|B|^2\right) & B_x B_y \\ B_y B_x & \left(B_y^2 - 1/2|B|^2\right) \end{bmatrix} \begin{bmatrix} l_x \\ l_y \end{bmatrix} . \tag{7.8}$$

Calculation of the Lorentz force from eqn (7.6) gives a value of 0.79 N for the motor with U-spring supports which is described in Section 5.2. The force-versus-current characteristic is linear, as is the characteristic based on experiments, and the thrust coefficient equals 3.95 N A^{-1} for each value of excitation.[15]

7.4.3 *Computation of the field and force using three-dimensional packages*

A three-dimensional finite-element package called TOSCA, which was developed at the Rutherford Appleton Laboratories and at Vector Fields Limited in the UK, has been used to calculate the fields in d.c. linear motors. It can also be used to calculate static forces; the calculated values can be compared with results obtained experimentally from motors which have the same specifications as the computer models.

The finite-element software package TOSCA can be used for the solution of magneto-static and electrostatic fields in three dimensions.[4] Problems are solved in three stages: SCARPIA, an interactive preprocessor, helps users to construct a finite-element mesh which is suitable for solution. Postprocessing of the results is achieved by using the interactive program POST TOSCA, which enables users to examine, graphically, the output from the solver.

The basic finite element used in models of square-armature d.c. linear motors is an isoparametric brick with eight nodes. This model has a total of

2350 nodes and 1800 elements. For greater accuracy, the air regions between the magnet poles and the armature are divided into quadratic elements rather than the linear elements which are used in other regions.

The TOSCA program uses material characteristics to relate the flux density and field intensity of all materials. For the yoke in the moving unit, it is defined in the first quadrant of the magnetization curve, whereas for permanent magnets the demagnetization curve in the second quadrant is used. The flux distribution[16] in every part of a square-armature motor has been calculated using the TOSCA program, and the result is shown in Fig. 7.4. Visual examination of the flux distribution will enable an engineer to modify the design within a short period of time.

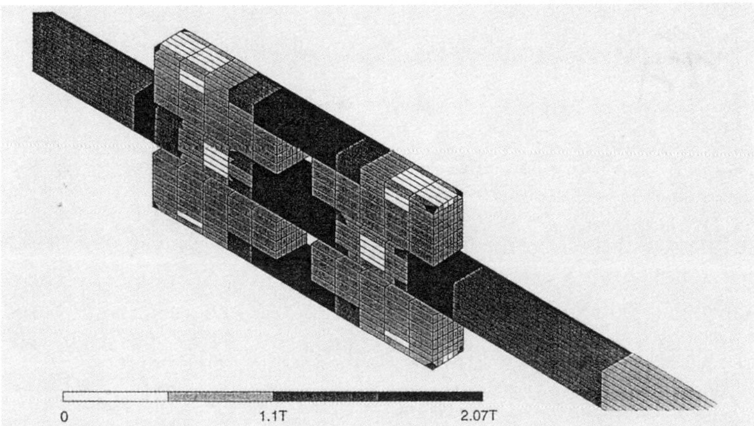

Fig. 7.4. The flux density distribution calculated throughout a square-armature motor.

The force has been calculated for a similar motor which has a circular armature. For this motor, the length of one armature winding section is 114 mm, the diameter of the armature core is 25.4 mm, the diameter of the armature conductor is 2.4 mm, and the armature current is 25 A. The armature-winding section is, therefore, modelled as a 114 mm long and 2.4 mm thick tube with an inner diameter of 25.4 mm, and a current density of 5.2×10^6 Am^{-2}. The tube is divided into 16 parts to calculate the force. The force developed in each part is found by taking the average of the vector cross products of the current density and the flux density at eight positions. The total force is then obtained by summing these forces. A model of an armature winding section used for the computation of the force is shown in Fig. 7.5.

7.5 Static analysis

Measured and ideal values of the static thrust, with the variation of the armature current for the motor with a square armature which is shown in Fig. 2.20, are illustrated in Fig. 7.6. The measured characteristics are straight lines as we would expect, but they do not pass through the origin as they ideally should do. They cross the armature current axis away from the origin because of various types of mechanical friction. These offset straight lines can be represented by the simple equation

$$F = \phi\, n\, (I_a - I_f), \tag{7.9}$$

where $\phi\, n\, I_f$ is the force required to overcome the friction.

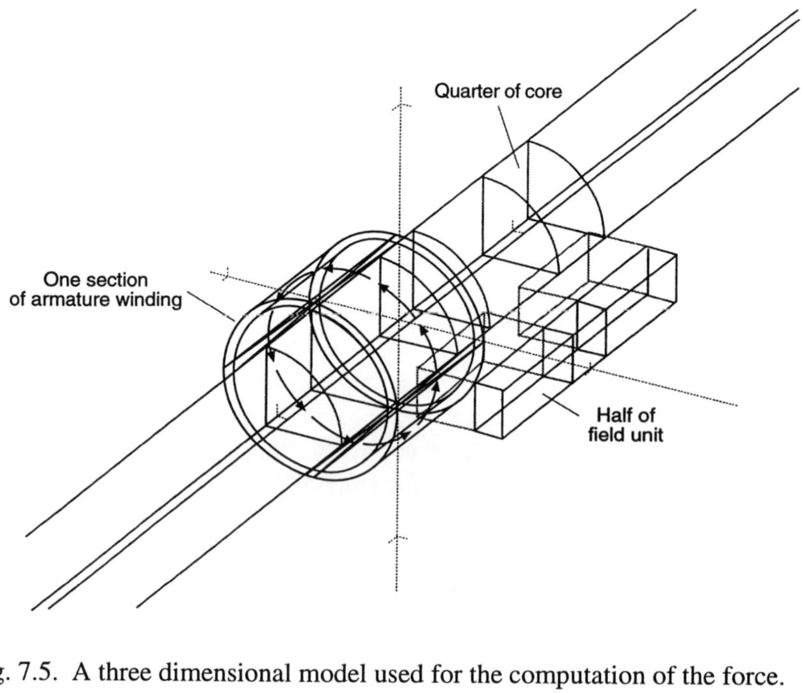

Fig. 7.5. A three dimensional model used for the computation of the force.

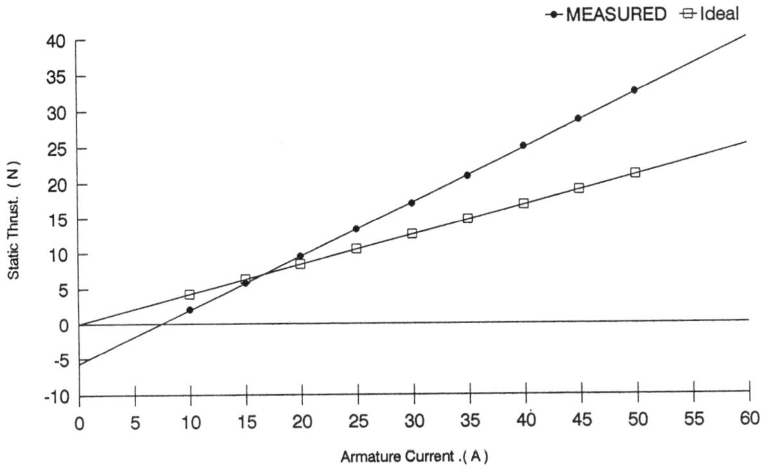

Fig. 7.6. Measured and ideal static thrusts.

The thrust was also measured in each of 12 multilayer coils of a motor as the armature current varied. This motor was designed to develop a force of 2 N for an armature current of 0.8 A. The thrust-versus-armature-current characteristics are shown for six coils in Fig. 7.7.

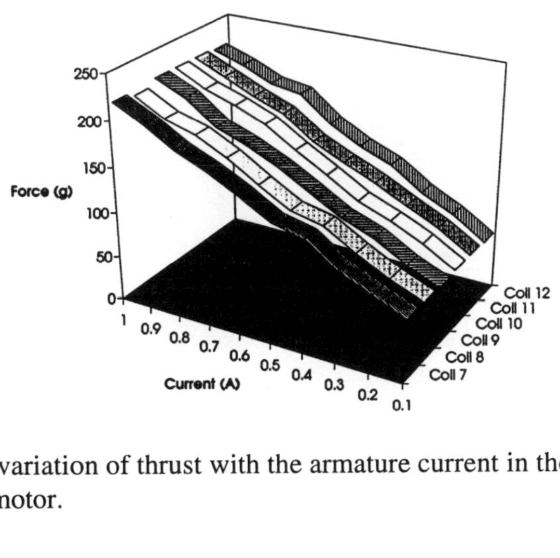

Fig. 7.7. The variation of thrust with the armature current in the multilayer coils of a motor.

The variation of axial flux flow in the core of the square armature motor is shown in Fig. 7.8(a). There is an increase in the magnitude of the armature flux, as expected, as the centre of the motor is approached. The reversal of the direction of flux near the outer edges of the moving unit is due to the leakage flux from the sides of the magnets. The calculated values of the axial flux density along the length of the motor are given in Fig. 7.8(b). If we multiply the flux density by the cross-sectional area of the armature core, 0.00 065 m^2, we can see that the results shown in Fig. 7.8(a) and (b) are comparable.

The density of radial flux reaching the armature from the magnet poles of the same motor, in magnitude and polarity can be measured with the aid of arrays of Hall-effect elements, as shown in Fig. 7.9(a). Knowledge of the radial flux at the armature surface enables the useful flux to be determined. It also gives information about the leakage flux. Figure 7.8(b) illustrates the calculated radial component of the flux density in the air gap between the armature winding and the magnet poles; the air gap is a critical region of the motor for energy transfer and the leakage field. The radial flux density reaches its peak in the air gap opposite the middle of the magnet poles, and it becomes zero in the air gap opposite the spacer which is placed between two magnets with opposite poles. In this case there is a noticeable discrepancy between the experimental and theoretical results.

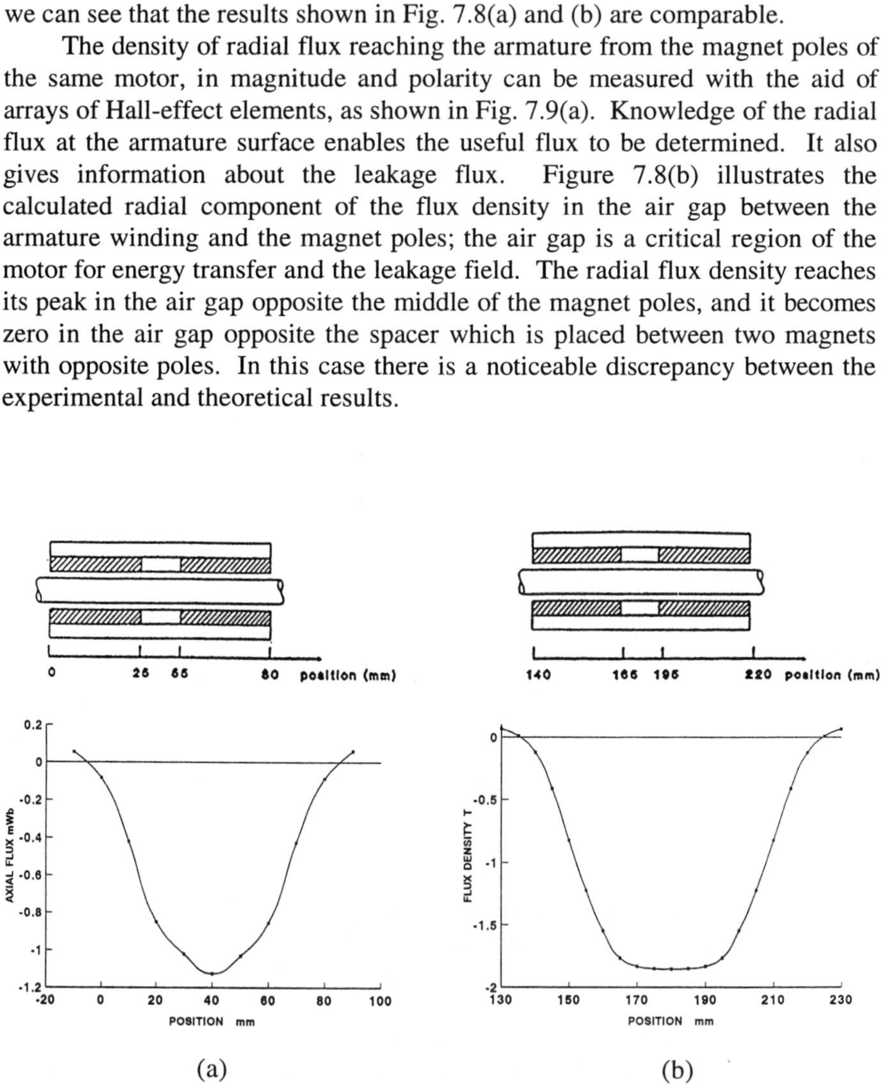

Fig. 7.8. (a) Measured and (b) computed axial fluxes along the armature.

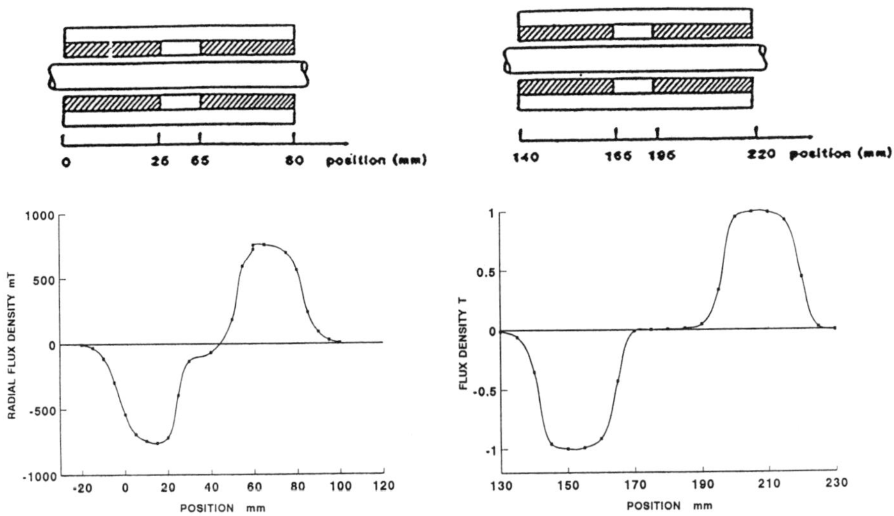

Fig. 7.9. (a) Measured and (b) computed radial flux densities.

The distribution of the modulus of flux density (BMOD) throughout the cross section of the motor can be calculated by the same two-dimensional software package. Studies of flux distribution usually concentrate on motor regions where the field is expected to change rapidly, in order to minimize computing time and the memory required. The density of the finite elements is, therefore, chosen to be largest in the armature core, in the yoke, and in the air gap between the permanent magnets and the armature winding. The modulus of flux density distribution throughout the cross section of the motor and in the surrounding air space is illustrated in Fig. 7.10. Observe that in Fig. 7.10 the areas close to the inner side surface of the magnets touching the yoke and the areas close to the armature immediately opposite the magnet poles have much higher flux densities than any other regions.

The flux through the yoke represents the total flux for the magnetic circuit of the motor. Numerical results, as shown in Fig. 7.11, obtained with the aid of the software package, indicate that the axial flux density increases as the observation points move towards the middle of the yoke. The axial flux density is negative at the ends of the model because of the leakage field in this area.

A useful plot of the calculated magnetic-vector-potential distribution throughout the cross section of the motor and in the air space is illustrated in Fig. 7.12. Lines of equipotential are drawn through the cross section of the

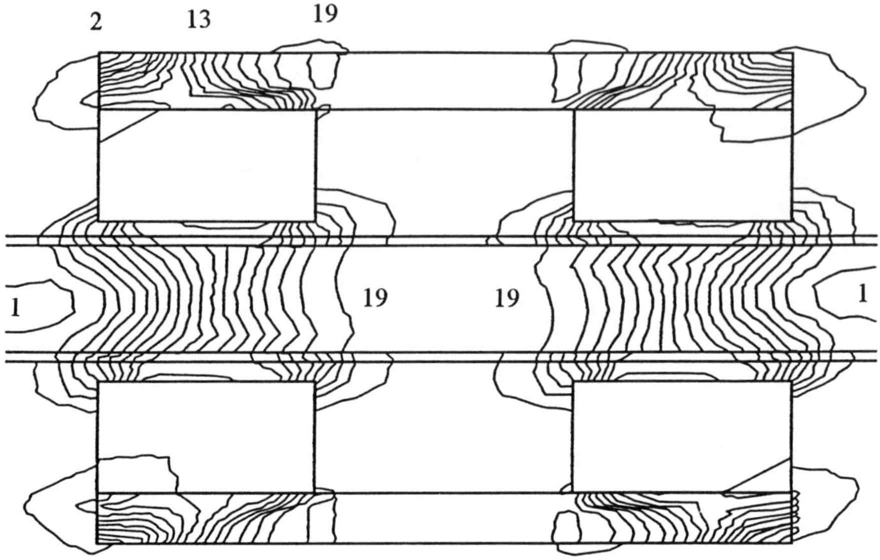

Minimum: 0.1 T, Maximum: 1.9 T, Interval: 0.1 T

Fig. 7.10. The modulus of the flux density in a motor with a square armature (contour 1: 0.1T and contour 19: 1.9T).

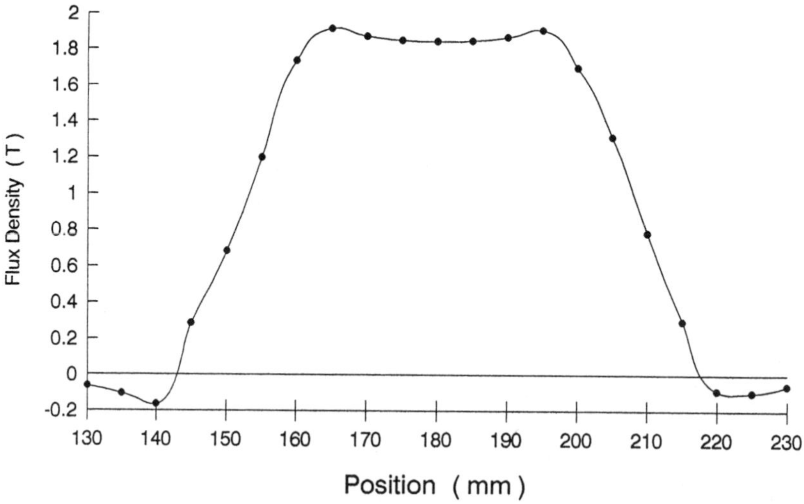

Fig. 7.11. The axial flux density along the yoke of a motor with a square armature.

motor. The kernel of the field distribution is located at the inner side surfaces of the magnet poles shown by contours marked 1 (–24 mWb m^{-1}) and 25 (+24 mWb m^{-1}). A uniform reduction in the magnetic potential is observed passing from the kernel of the field along the radial lines. Figure 7.12 shows that there is a high concentration of the magnetic potential in the armature core near the middle of the model and at the rear of the yoke behind the magnet poles. Most of the flux reaches the armature core by crossing the air gap between the armature winding and the magnet poles, and since the flux cuts the conductors carrying the armature current at right angles the motor develops a force.

The flux distribution within the armature beyond the moving unit and in the free space surrounding the motor both confirm that some of the leakage fluxes does not cross the air gaps beneath the magnet poles. The major flow path of these fluxes is via the ferromagnetic material of the moving unit. The most significant source of the leakage flux is in the interpolar region. A reduction of the interpolar path length by a decrease in the separation length between two opposite magnet poles results in an increase in the leakage component. From the edges of the moving unit the leakage flux follows a long route through the air space. Again from different parts of the machine the leakage flux gives rise to a nonuniform distribution of flux about the armature periphery, hence part of the armature winding lies in a region of low flux density, and this part makes a reduced contribution to the force which is developed by the motor.

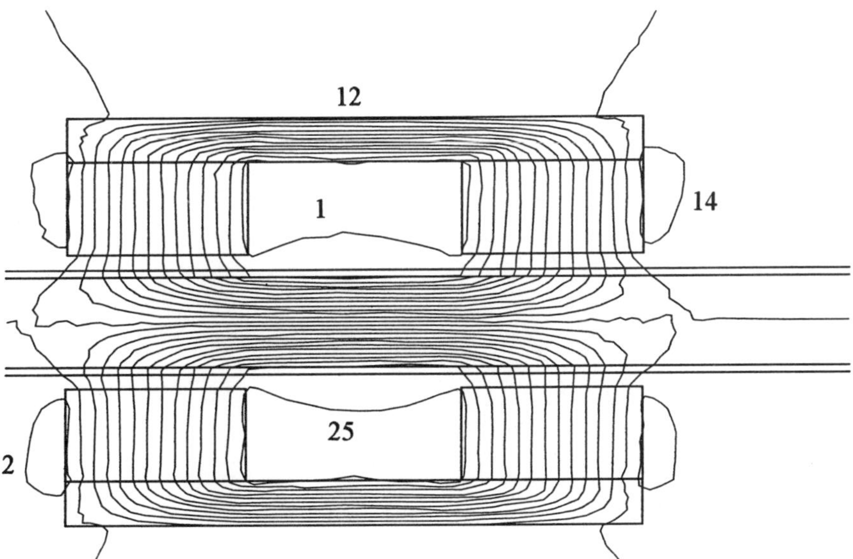

Fig.7.12. Equipotential lines in a motor.

Once the magnetic-vector-potential distribution throughout the machine is known, the force developed by the machine can be obtained since this distribution enables the radial component of the flux density in the air gap between the armature section and the magnet pole to be calculated. The total force is obtained by calculating the mean value of the radial component of flux density in the air gap over the entire length of the armature section.[2] The force which is thus obtained is based on the flux density profiles, and it represents the total developed force at standstill. The force–current characteristic may be calculated for the same range of excitation as for the experiments. There are two possible approaches to this. The first requires a separate problem definition for each excitation. The second uses the scaling factor which is offered by a solver. The only disadvantage of the latter method is the difficulty in handling: increased processing time and the bigger file which results but on the other hand, there is only one data loading, mesh generation and sparsity-pattern formation.

7.6 Dynamic characteristics

The velocity of the square-armature d.c. linear motor described earlier can be measured for variations in the armature current using the technique described

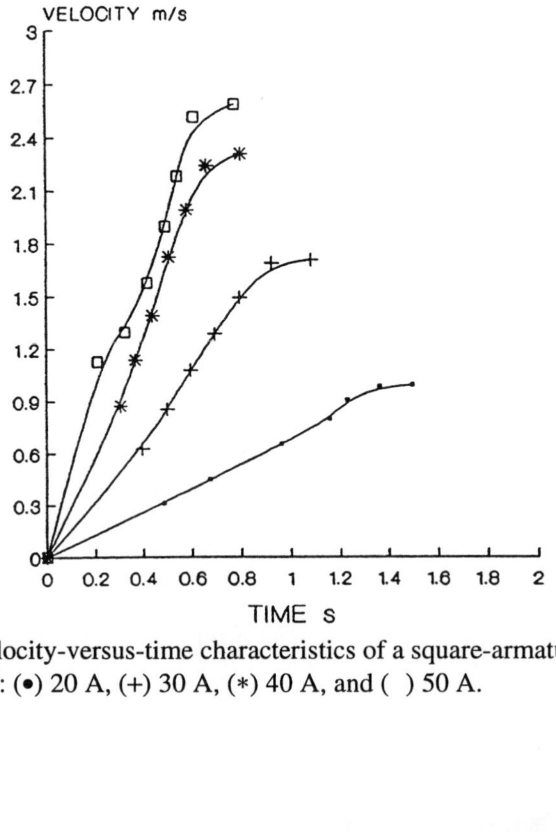

Fig. 7.13. Velocity-versus-time characteristics of a square-armature motor for currents of: (•) 20 A, (+) 30 A, (*) 40 A, and ( ) 50 A.

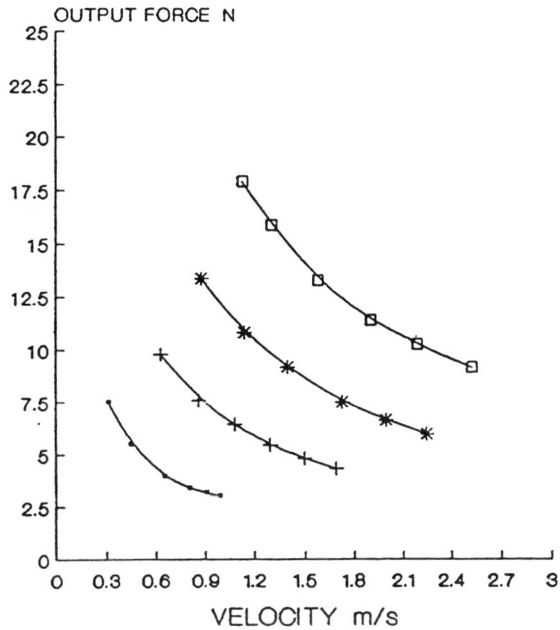

Fig. 7.14. Output-force-versus-velocity characteristics of a square-armature motor for currents of: (•) 20 A, (+) 30 A, (*) 40 A, and ( ) 50 A.

in Section 7.3.2. The air gap between the armature and the magnet poles is kept fixed at 4 mm. This is comparatively larger than that in the rotary counterpart because of the long and linear track. Figure 7.13 shows the velocity-versus-time characteristic for this motor. Observe that the velocity increases linearly with time for a long period, because of the friction generated by the support and the guide system, before it reaches a steady state. This happens at all armature excitations. At lower excitations the motor not only has lower steady speeds, but it also takes longer time to reach these speeds.

The output force of this motor can be determined from a knowledge of the acceleration and the mass of the moving unit. Figure 7.14 illustrates the variation in the output force with velocity at different excitations. Observe that in Fig. 7.13 the acceleration gradually decreases as the velocity increases. Therefore, the output-force-versus-velocity characteristic should be expected to have the shape it has in Fig.7.14.

Both d.c. and a.c. linear motors produce considerably less output power, and have much lower efficiencies and inferior power-to-weight ratios than their rotary counterparts. The low efficiency of d.c. linear motors is almost entirely

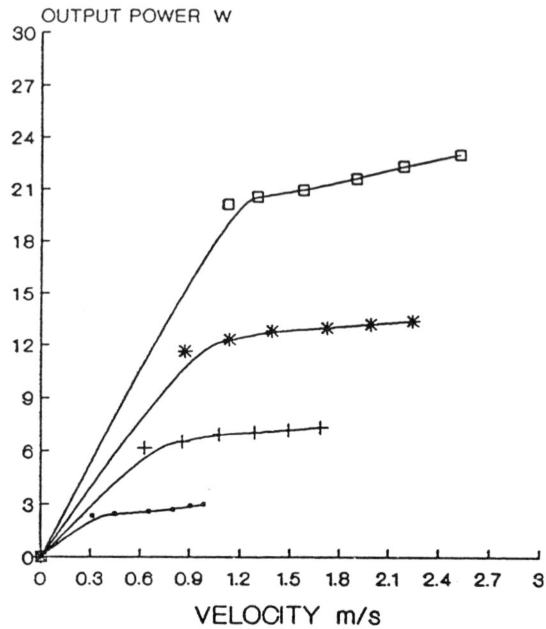

Fig. 7.15. Output-power-versus-velocity characteristics of a square-armature motor for currents of: (•) 20 A, (+) 30 A, (∗) 40 A, and ( ) 50 A.

due to the fact that they operate with a large air gap, which results in a relatively high leakage flux, and because of the friction from the support and guidance system. The output-power-versus-velocity characteristic and the efficiency-versus-velocity characteristic for the square armature motor for different excitations are shown in Figs 7.15 and 7.16, respectively.

7.7 Magnetic-circuit parameters

The equivalent magnetic circuit for the square-armature d.c. linear motor is shown in Fig. 7.17. By drawing the permeance lines on the flux against the m.m.f characteristics, the values of the useful reluctance, $\Re_u$, the leakage reluctance, $\Re_l$, the open-circuit reluctance, $\Re_o$, and the open-circuit m.m.f., F_o, can be determined so that the magnetic circuit can be established for the motor.

The flux-versus-m.m.f. characteristic is obtained by transforming the demagnetization curve of the permanent magnet which is used in the motor. Multiplying the values of the field strength, H, by the magnet length gives values for the corresponding m.m.f., and multiplying the values of the flux

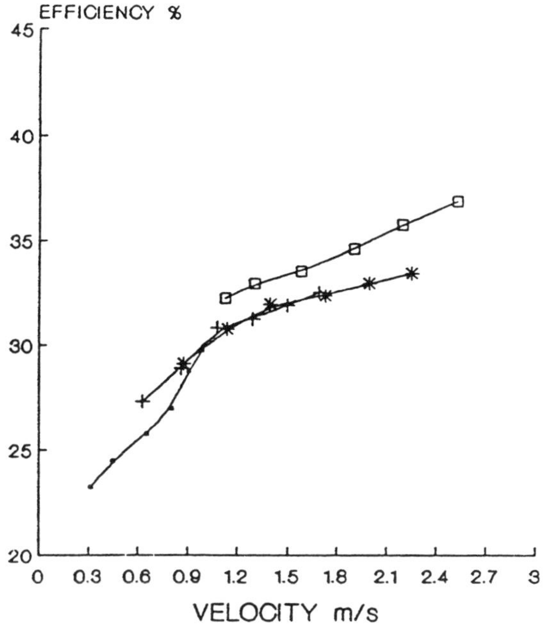

Fig. 7.16 Efficiency-versus-velocity characterics of a square-armature motor for currents of: (•) 20 A, (+) 30 A, (*) 40 A, and ( ) 50 A.

density, B, by the magnet pole area gives values for the corresponding flux, ϕ. The flux-versus-m.m.f. characteristic for the motor is shown in Fig. 7.18; ϕ_r is the remanent flux, and F_c is the coercive m.m.f.

The total flux in the moving unit of the motor is measured with a search coil and a demagnetizing coil wrapped round the steel yoke and with a ballistic galvanometer. The measuring technique[13] was described in Section 7.3.7. The total flux, ϕ_t, emanating from the magnets divides into the radial flux, which produces the motion (this is known as the useful flux, ϕ_u), and the leakage flux, ϕ_l. The useful flux crosses the air gap, and it links the armature winding. It can be measured by the technique described in Section 7.3.5. The principal source of the leakage flux is between two adjacent but opposite magnet poles, shown in Fig. 7.19. The useful flux, the total flux, and the magnet pole area are 1.03 mWb, 1.54 mWb and 625 mm^2, respectively, for this motor.

The leakage and useful flux are represented by the lines ab and bc, respectively, in Fig. 7.18. A straight line parallel to the m.m.f.-axis is drawn from the point c which cuts the ϕ–m.m.f. curve at point d. A line de, which is perpendicular to the m.m.f.-axis, represents the total flux ϕ_t emanating from the magnets, while the line cd represents the m.m.f. for both the leakage path, F_l

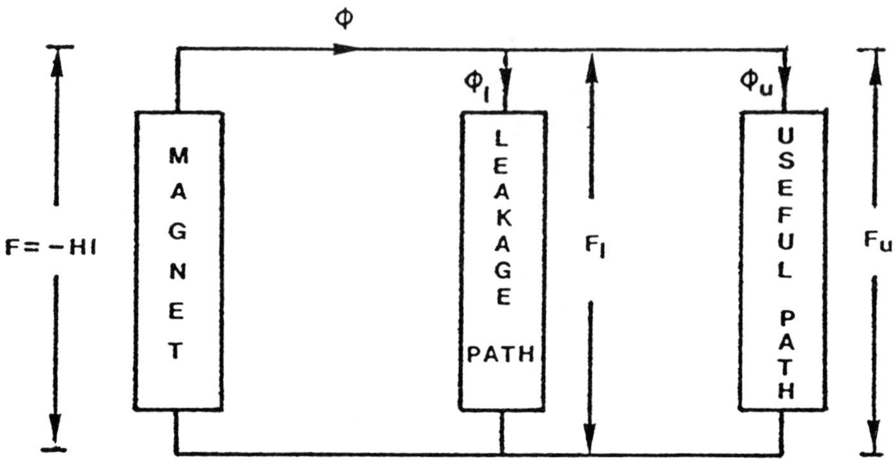

Fig. 7.17. An equivalent magnetic circuit.

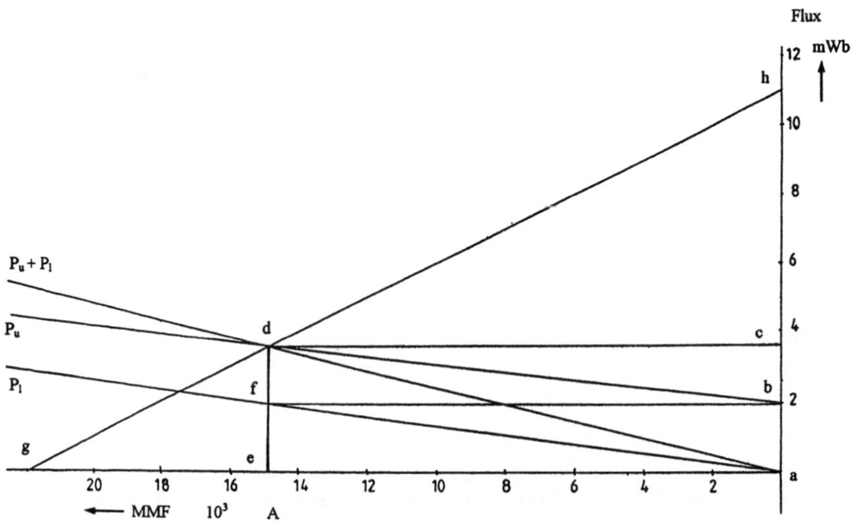

Fig. 7.18. Flux-versus-m.m.f.-curve and permeance lines.

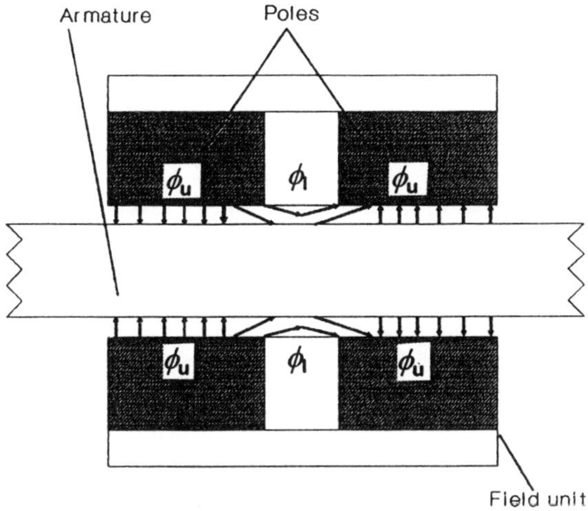

Fig. 7.19. Useful and leakage flux paths.

and the useful path, F_u. A line parallel to the m.m.f.-axis is drawn from point b to cut the line de at the point f. Using simple geometry, it can be seen that, in the rectangle acde, the lines ef and ab and the lines fd and bc are equal, hence the lines ef and fd represent the leakage and useful fluxes respectively.

From basic magnetic-circuit theory,

$$\phi = FP = \frac{F}{\mathfrak{R}}, \tag{7.10}$$

where ϕ is the flux (in webers), F is the m.m.f. (in amperes), P is the magnetic permeance (in Wb A^{-1}), and $\mathfrak{R}$ is the reluctance (in A Wb^{-1}).

Hence the lines bd and af represent the permeance lines with slopes P_u and P_l, and the line ad represents the total permeance line with the slope P_u+P_l due to the useful and leakage fluxes. The reluctances $\mathfrak{R}_u$ and $\mathfrak{R}_l$ are the reciprocals of P_u and P_l respectively.

A line passing through the point d, with a slope equal to the slope of the tangent of the ϕ–m.m.f. curve at the point ϕ_r, represents the recoil line, and its extension intersects the ϕ-axis and the m.m.f. axis at the points h and g respectively. The line ag represents the open-circuit m.m.f. F_o, while the line ah represents the flux at a zero m.m.f., ϕ_0. The working point of the magnet is on the recoil line, and the slope of this line gives the reciprocal of the open-circuit reluctance, $\mathfrak{R}_o$. Table 7.1 gives the values of various reluctances.

Table 7.1 Reluctances of the magnetic circuit of the motor

$\mathfrak{R}_l$ (A Wb^{-1})	$\mathfrak{R}_u$ (A Wb^{-1})	$\mathfrak{R}_o$ (A Wb^{-1})	$\mathfrak{R}$ (AWb^{-1})	F_o (A)
18.85	9.41	7.97	14.25	21.93

The resultant reluctance of the magnetic circuit in Fig. 7.17 can be calculated using

$$\mathfrak{R} = \mathfrak{R}_o + \frac{\mathfrak{R}_l \times \mathfrak{R}_u}{\mathfrak{R}_l + \mathfrak{R}_u} . \qquad (7.11)$$

The resistance and the inductance of the armature winding, R_a and L_a, may be measured using a Wheatstone bridge and a d.c. bridge circuit respectively. The electrical time constant, τ_e, is the ratio of the armature inductance to

Table 7.2 Electrical parameters for a square-armature motor

R_a (Ω)	L_a (H)	τ_a (s)
0.06	1.91	32

Table 7.3 Parameters of a square-armature motor

Load mass (kg)	Current (A)	K (Wb m^{-1})	v_o (m s^{-1})	F_{eff} (Ns m^{-1})	τ_m (s)	f (Ns m^{-1})
0.25	20	0.82	0.9	15.1	0.6	8.6
0.50	30	0.82	1.7	12.0	0.7	5.6
0.50	40	0.82	2.1	13.6	0.7	7.2
0.75	40	0.82	2.0	13.2	0.7	6.7
0.75	50	0.82	2.4	14.2	0.6	7.8
1.00	50	0.82	2.1	15.1	0.6	8.7

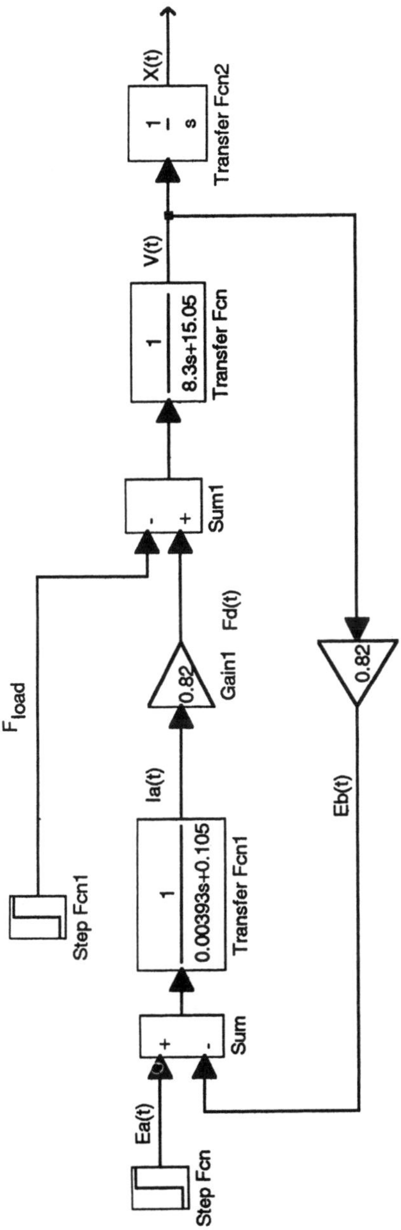

Fig.7.20. Calculation of transfer function of a d.c. linear motor with a square armature.

the armature resistance, and it is given in Table 7.2 for the square-armature d.c. linear motor.

The apparatus shown in Fig. 4.9 may be used to measure the velocity of a motor under a variety of load and excitation conditions. The velocity-versus-time characteristics of the square armature d.c. linear motor with a mass of 8.3 kg have been used to obtain the steady-state velocity for different loads and armature excitations. The coefficient K is the product of the magnet flux, ϕ, and the number of turns per metre of armature length, n. By substituting the values of the steady-state velocity, v_o, and the values of the coefficient K in eqn (4.73), the time constant of the motor, τ_m, and the effective frictional coefficient, F_{eff}, may be determined. Table 7.3 shows the values of v_o, K, τ_m, and F_{eff} for the d.c. linear motor with the square armature for loads ranging from 0.25 to 1.0 kg and armature currents varying from 20 to 50 A.

The parameters presented in Table 7.3 for various current values were substituted in the transfer function for the square armature d.c. linear motor, as shown in Fig. 7.20. The program SIMULINK[73] was used to plot the velocity-versus-time characteristics for the square-armature motor under various load conditions in Fig. 7.21.

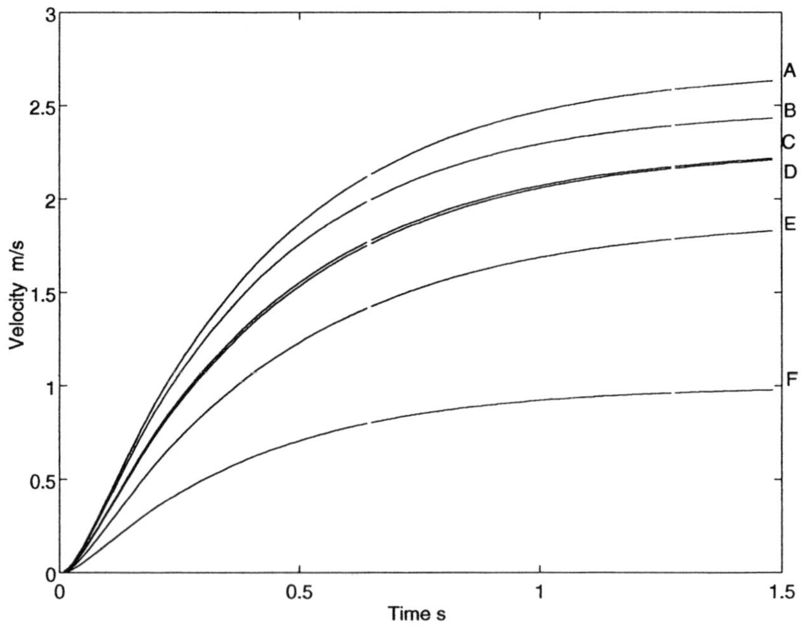

Fig. 7.21. Simulated-velocity-versus-time characteristics for loads of: (A) 50A, 0.75 kg; (B) 50 A, 1 kg; (C) 40 A, 0.5 kg; (D) 40 A, 0.75 kg; (E) 30 A, 0.5 kg, and (F) 20 A, 0.25 kg.

This chapter should enable designers to obtain various experimental and computational results for a motor, and then to use these results to improve the design of a motor with the aid of the relevant equations from Chapter 4.

8

APPLICATIONS

There are advantages in light and heavy industries in using linear rather than rotary motors for linear drives if the position and speed of these motors can be controlled precisely. Linear motors remove the necessity for gears, belts, etc., to transfer the rotary motion into linear motion. Thus they reduce the cost of the drive system, and also the space required for it. Direct-current linear motors offer an impressive combination of precision, power, and speed. They can produce a thrust of up to 1 kN, accelerate loads to speeds in excess of 1.5 m s^{-1}, and then position the loads to submicrometre accuracies.[76] This type of motor is, however, not easily adaptable to a variety of application demands (as rotary motors are), since gearing is impracticable. Therefore, in most linear-drive applications, requirements such as stroke-length, linear-velocity, acceleration and deceleration levels, etc., need a motor which is uniquely designed for a particular application. In addition, the decision as to which type of moving magnet, moving armature or moving coil is best for a given application has to be made after the motion requirements are known.[66] The major disadvantage of d.c. linear motors is their need for commutators and brushes, but this drawback is easily avoided by using electronic commutation. The following list gives the main areas in which d.c. linear motors are applied:

(1) materials handling in large and small industries,
(2) long-stroke positioning machines,
(3) positioning systems for robots,
(4) medical-check-up equipment,
(5) detector head drives in industry,
(6) audio and lighting mixing decks,
(7) clean-room transportation,
(8) artificial hearts,
(9) textile machines, and
(10) replacing hydraulic/pneumatic linear drives.

8.1 Doors and materials handling

A d.c. linear motor develops a large output force at zero speed. The force decreases as the speed increases. This characteristic makes it most useful in applications where relatively low speeds are required. One of the major applications is to power sliding and folding doors, because these motors theoretically provide a simple mechanical arrangement which gives a long life, and low maintenance costs. Another important application is in the transportation of materials in factories. A conveyor for assembly or packaging work can be developed using one or more d.c. linear motors.

The square-armature d.c. linear motor with neodymium–iron–boron magnets[3] described in Chapter 2 with a track length of 4 m, and a 15 m long d.c. linear motor with ferrite permanent magnets[13] have both proved their ability to transport materials as suggested above. Figure 8.1 shows how a d.c. linear motor can be used to move doors. Speed control of sliding and folding doors powered by linear motors is essential to prevent slamming or over-fast operation. Doors are accelerated at full thrust up to the set speed, then a lower current is supplied to the armature that is sufficient to maintain that speed. As the door approaches the stopping point, the armature current is reversed to slow the door down, then the current is gently reduced to shut the doors completely. A typical elevator door can be driven by a 400 W machine, and the fact that its efficiency is low is generally of secondary importance. This system has several advantages over rotary-motor drives, including reduced costs. The electrical supply to the motor can be obtained by using a centre-tapped transformer and two diodes.

Materials can be placed in flat trolleys, which are supported by the top of the moving unit of a d.c. linear motor, and transported. These trolleys carrying materials for assembly or packaging work can be moved at low speeds, and stopped at various predetermined positions for chosen periods of time, either manually or with the help of computer instructions. A single armature core can be used for the whole assembly line with a number of worktops, each of which is carried by an individual moving unit.

Other applications of d.c. linear motors in this field can be found in laser welding and cutting materials, coil winding, and also in the textile industry. Linear motors are preferred in coil winders for their precise position control, and for their mechanical simplicity over a long travel length. In the textile industry these motors can speedily cut broad cloth with blades. A moving object can also be cut by a saw riding on a linear motor which is driven at a synchronous speed, with the aid of sensors and a control circuit.

8.2 Positioning systems

Dascalescu *et al.*[25] have studied several types of electrical machine in order to achieve accurate and effective position control of the electrodes in corona electrostatic separators. They found that linear a.c. induction motors are less adequate in position-control duties than linear d.c. motors. Eventually they chose the motor conceived by Basak and Overshott,[17] in which the field excitation is provided by ferrite-permanent-magnet blocks.

The position of the electrodes significantly influences the electrostatic separation. Each electrode-positioning control parameter necessitates an adequate actuating device (see Fig. 8.2). Direct-current linear motors can easily control the distances, ranging from 50 mm to 150 mm, between the surface of the drum and the rotation axis of the electrodes. Stainless-steel electrodes 25 mm in diameter and 750 mm in length are moved at a speed greater than 1 m s^{-1}, and the precision of positioning is better than 0.1 mm.

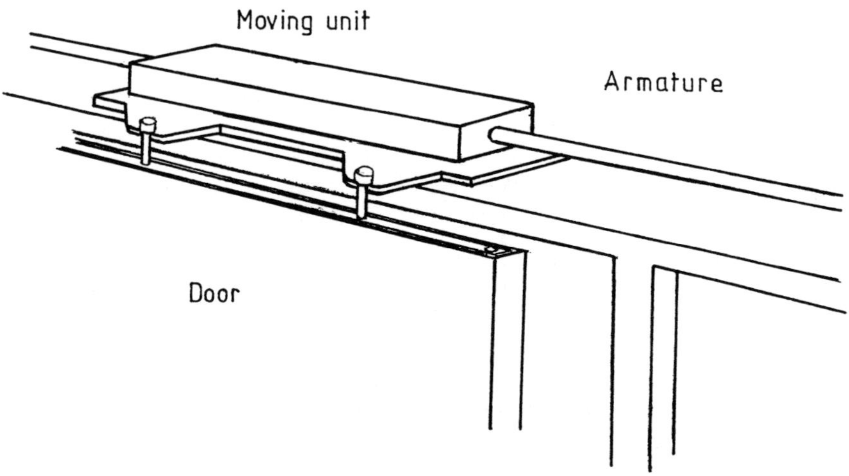

Fig. 8.1. Sliding doors.

Ferrite field d.c. linear motors (such as the motor shown in Fig. 8.3) meet the above requirements. They consist of a stationary armature and a moving-field system. The armature is made of a 200 mm long mild-steel bar 34 mm in diameter, with a single layer of enamelled copper wire 0.9 mm in diameter closely wound on it. Removal of the wire insulation from the top surface of the winding, along which the motor runs, provides the motor with a commutator. The armature current can be at least ten times the rated current with no requirement for cooling, since the armature core provides an excellent heat

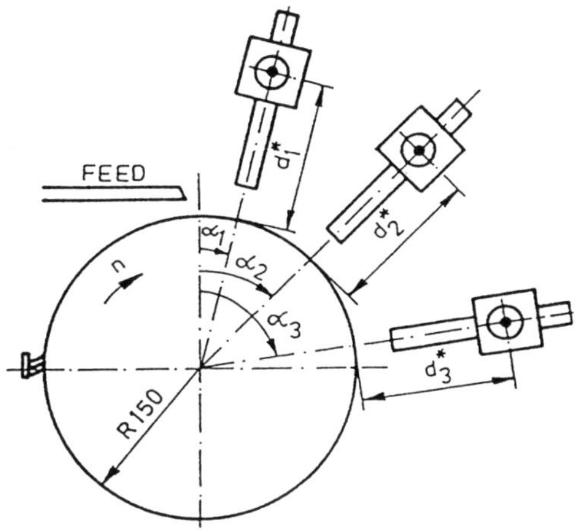

Fig. 8.2. The electrode-positioning parameters.

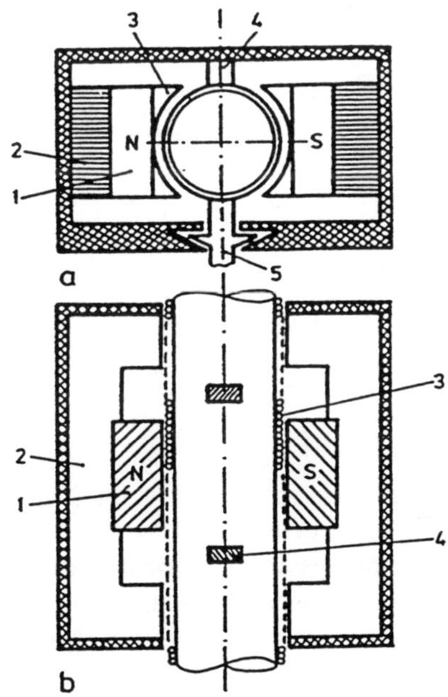

Fig. 8.3. (a) Transverse and (b) longitudinal sections of a d.c. linear motor used for an electrostatic separator: (1) magnet blocks; (2) silicon–iron laminations; (3) the armature winding; (4). the brush; and (5).the armature support.

sink, and the position of the brushes ensures that the winding is only subjected to the current during the short period of the traverse of the machine. The distance between the brushes is 27 mm, so 30 turns of winding can be accomodated in the active zone of the motor. The flux density is 0.22 T in the 1.5 mm long air gap.

The field system consists of yokes fabricated from silicon–iron laminations, and ferrite-permanent-magnet blocks. High-current carbon brushes, to energize the armature winding, are located on either side of the active poles. Thus the radial magnetic field reacts with the circumferential current flow in the armature winding, developing a force which enables the carriage to move. Figure 8.4 shows the computed speed–force characteristic of Dascalescu's motor.

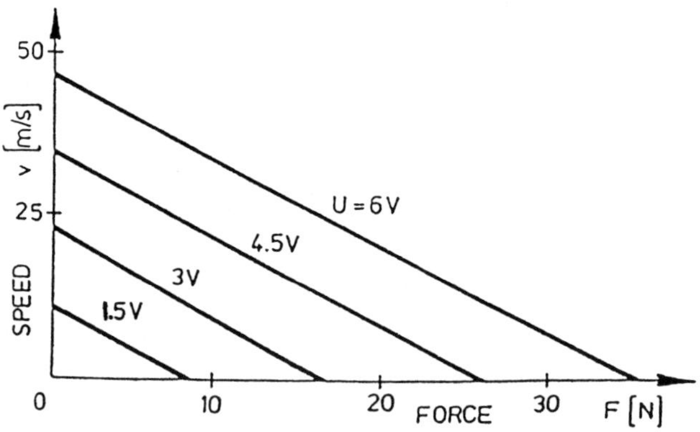

Fig. 8.4. Computed force–speed characteristics for the motor of an electrostatic separator.

Direct-current linear motors can also be used in the focusing mechanism of cameras. Again the cameras themselves can be scanned with the aid of these motors. The Toyota Motor company has developed a moving-coil-type of d.c. linear motor to operate the window curtains in their cars.[59] A stroke of 1200 mm is achieved with a thrust of 2.8 N at 4.8 A and 12 V. Ferrite magnets are used to provide the excitation.

8.3 Audio mixing decks

In many applications, such as studios, it may be necessary to move one or more sliders, to control lighting or audio equipment, to predetermined positions.

This has previously been done by using faders with d.c. rotary motors; this required gearboxes, moving belts, and ancillary driving equipment, which are sources of inaccurate drives. Now industries have developed faders which make use of d.c., linear, moving-coil motors.

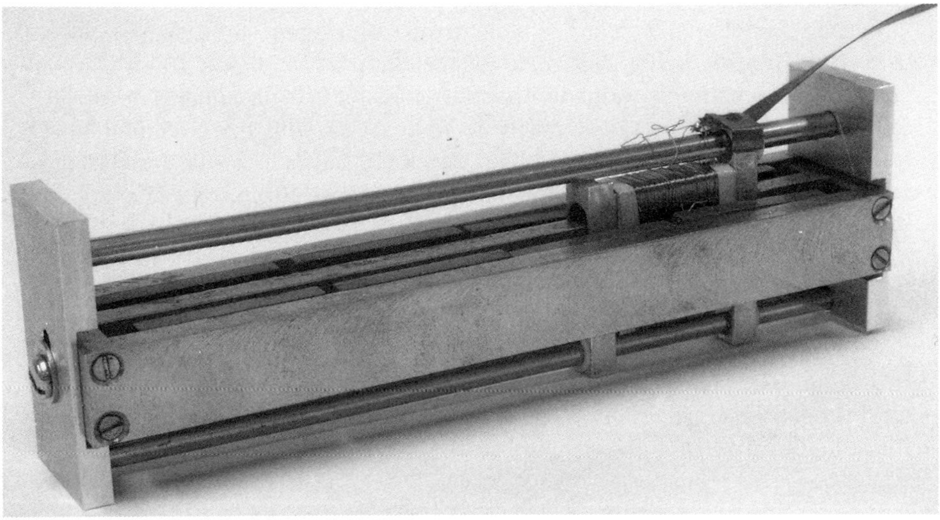

Fig. 8.5. A slider used for an audio mixing or fading system.

Figure 8.5 shows the prototype of a slider developed by a studio-equipment manufacturer using the principles described in Section 2.5. A typical number of sliders in an audio deck is 30, and this may limit the maximum current available for one slider mechanism to around 30 mA.[18] Because of this, and because of the characteristics of the magnets which are used, and also because of the limited width of audio decks, multilayer coils have to be wound on the armature in order to produce the required thrust.

8.4 Multiaxis drive systems

Multiaxis drives have a variety of applications both in light and heavy industry. *X–y* chart recorders, which are very popular in research and teaching laboratories, can be built by using two low-power, d.c. linear motors, and a similar arrangement but with more powerful motors is employed for positioning tools or objects for drilling, milling, or punching; this drive system can be used in medical-scanning applications.

8.4.1 *Chart recorders*

The principle of operation of the linear motor used in the Foster-Cambridge Clearspan P100L electronic recorder was described in Chapter 1 and the motor is shown in Fig. 1.21.

Another chart recorder[79] built with two moving-magnet-type d.c. linear motors is shown in Fig. 8.6. The armature winding of one motor has 20 sections, each with five layers of 0.2 mm diameter enamelled copper wire. The other motor has an armature with 30 sections of winding. The permanent magnets used in the sliders are rare-earth neodymium–iron–boron. In the y-axis (with a smaller travel length) motor the static thrust varies within 5 per cent from coil to coil, as can be observed in the characteristic shown in Fig. 8.7 for an armature current of 1 A. Due to the friction experienced by the slider, the difference between the theoretical and practical thrust is around 12 per cent. A similar observation can be made for the x-axis (which has a longer travel length) motor.

Figure 8.8 shows a diagram of a pen recorder, described by Wakiwaka *et al.*,[80] and a block diagram of its servomechanism. The open-loop transfer function G is given by

$$G = \frac{K_\mathrm{A} K_\mathrm{P} \mathrm{e}^{-st_\mathrm{L}}}{sNlB(s^2\tau_\mathrm{e}\tau_\mathrm{m} + s\tau_\mathrm{m} + 1)} \tag{8.1}$$

where K_A is the coefficient of the amplifier, K_P is the sensitivity of the potentiometer with units of $\mathrm{V\,m^{-1}}$, t_L is the delay of the potentiometer and the time lost due to free play, N is the number of turns, l is the length of coil affected by the flux density, B is the flux density acting on the coil, and I is the coil current. τ_e and τ_m are the electrical and motor time constants respectively. Note that

$$\tau_\mathrm{e} = L/R \quad , \quad \tau_\mathrm{m} = \frac{(M_c + M_\mathrm{L})R}{(NlB)^2}$$

where M_c is the mass of the coil, M_L is the mass of the load, and R is the resistance of the coil.

8.4.2 *Industrial multiaxis carriers*

In industry there is always a need to move machining tools to the required position of the product. Sometimes it becomes necessary to inspect and measure the parameters of the finished product. For these manoeuvres

Fig. 8.6. A plan view of a multiaxis carrier.

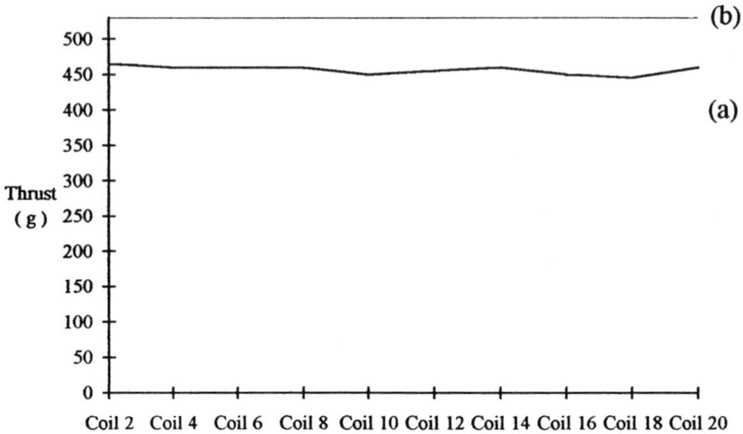

Fig. 8.7. The thrust of a *y*-axis motor for a current of 1 A: (a) measured value, and (b) the theoretical maximum value.

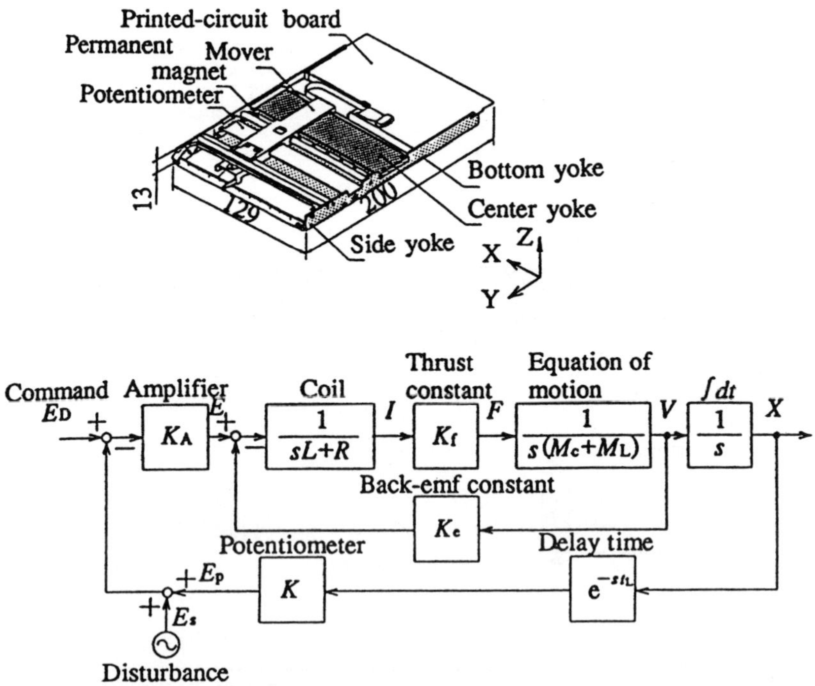

Fig. 8.8. A pen recorder (dimensions are in mm) and a block diagram of its servomechanism.

multiaxis carriers which are similar to chart recorders can be used. Other applications in light industries include printed-board multi-hole drilling, drafting, and plotting.

8.5 Artificial heart pumps

The blood-circulation system of living bodies is operated by two pumps; the heart pump and the peripheral pump. A blood-circulation simulator has been proposed by Yamada *et al.*[84] They used d.c. linear motors to construct the two pumps for the simulator, and they verified its applicability. Figure 8.9 shows the system configuration of the blood-circulation simulator. Blood is circulated by pumps A and B in the direction indicated by the arrows. Each pressure chamber is provided with a check valve, and the open reservoirs correspond to the atrium. Closed tanks 1 and 2 hold air layers in their upper sections to simulate lungs and peripheral blood vessels. The gate valves correspond to various pressures.

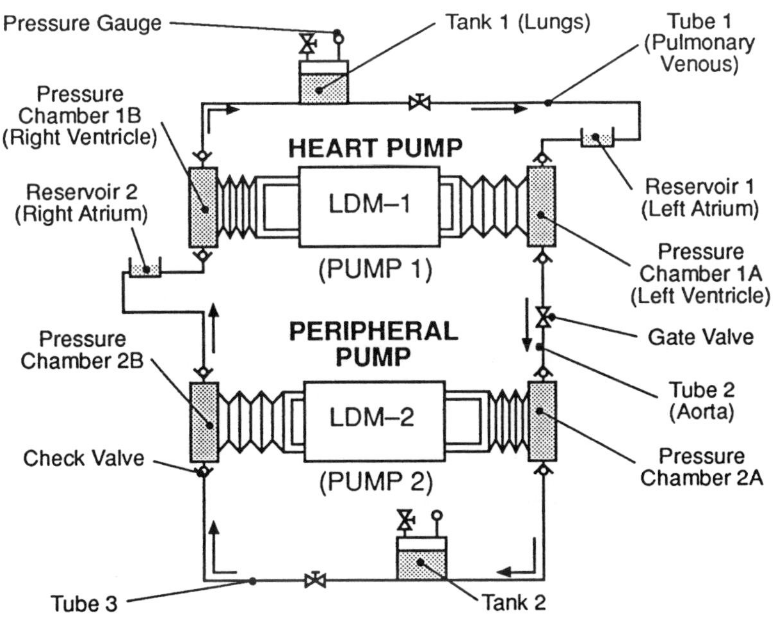

Fig. 8.9. A system configuration of a blood-circulation simulator.

The motors used for the simulator are of the moving-magnet type. The stator of each of the motors comprises a core, four coils, aluminium-roller-bearing guides, and a side plate. Each moving unit holds two pieces of rare-earth plastic magnets ($50\times35\times5$ mm^3) with the magnetic poles facing each other on either side of the core.

The static-thrust-versus-displacement characteristic is shown in Fig. 8.10 as a function of the exciting current for the motors used in the blood-circulation simulator. The thrust is 27 N when the exciting current is 4 A. The characteristic is approximately flat within the 30 mm stroke and it is, therefore, suitable for the simulator.

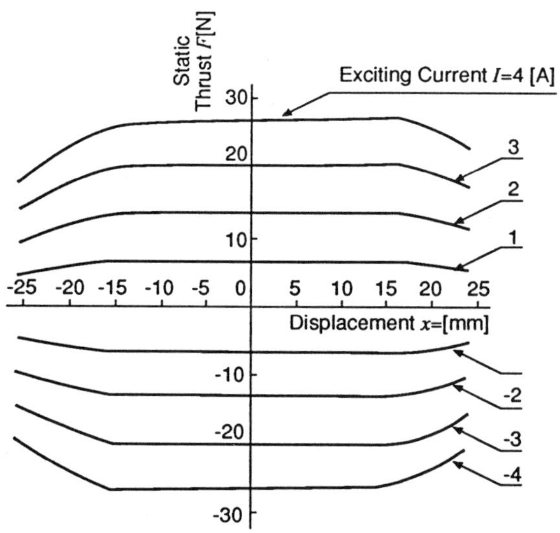

Fig. 8.10. Static-thrust-versus-displacement characteristics for various excitations.

8.6 Loudspeaker systems

In conventional loudspeakers the linear displacement is physically restricted by the suspension of the diaphragm. In order to deliver an adequate amount of low-frequency acoustic energy, conventional loudspeakers usually have a large diaphragm diameter; however, in a new system the power can be increased by increasing the linear displacement of the vibrating part, and the distortion due to the suspension can also be eliminated. The basic electromagnetic parts of the new system[55] are in the cylindrical configuration which is shown in

Fig. 8.11. The vibrating part is a piston made from composite materials inlaid with a neodymium–iron–boron permanent magnet. The coils are wound on the stator core. The interaction between the field generated by the permanent magnet, and the current-carrying coils produces the force. The motion of the piston and the electromagnetic levitation required to give a contact-free system are controlled by energizing the respective coils using electronic switching. An

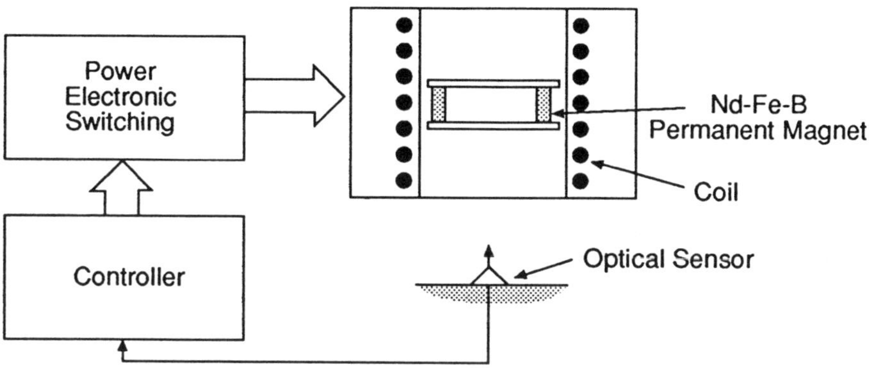

Fig. 8.11. Loudspeaker systems.

optical sensor at the bottom of the speaker detects the location of the vibrating piston, and it compares this with the desired location of the input signals. Any error signal which is generated then controls the current in the coils for the appropriate electromagnetic levitational force and for the required displacement.

8.7 Concluding remarks

The current available and the force demanded dictate the sensitivity required of the motors. In general, the sensitivity, that is, the output-force-to-current ratio of the motor, may be increased by increasing the motor width, thereby increasing the conductor length. Thus linear motors offer motion-control designers the option of generating a given force with less current if space is available for an armature with a larger cross-sectional area.

A few trade-offs can be made to provide a customized motor, but designers must know which parameters from those listed below are fixed by the application and which are flexible:

(1) maximum stroke length,
(2) total load inertia,
(3) frictional force,
(4) peak force,
(5) available package dimension
(6) supply voltage and current, and
(7) maximum speed.

With the proper specifications a d.c. linear motor can be custom designed to provide servo systems with the fastest and most accurate prime mover that is currently available.

APPENDIX A

A list of d.c.-linear-motor manufacturers

Manufacturer	Address	Product
Astrosyn International Technology plc	The Old Courthouse, New Road Avenue, Chatham, Kent ME4 4QJ	D.c. linear stepping motors
BSL Engineering Ltd	Claverton Court, Claverton Road, Wythenshawe, Manchester M23 9NE	D.c. linear stepping motors
Clark Components Company	Station Road, Brompton-on-Swale, North Yorkshire DL10 7SE	D.c. linear stepping motors
Laser Lines Ltd	Beaumont Close, Banbury, Oxfordshire OX16 7TQ	D.c. linear motors, brushless d.c. linear motors (short stroke)
Linear Drives Ltd	Brook Road Industrial Centre, Rayleigh, Essex SS6 7XL	D.c. linear motors
Unimatic Engineers Ltd	122 Granville Road, Cricklewood, London NW2 2LN	D.c. linear stepping motors, bearings

APPENDIX B

Products of Linear Drives Ltd

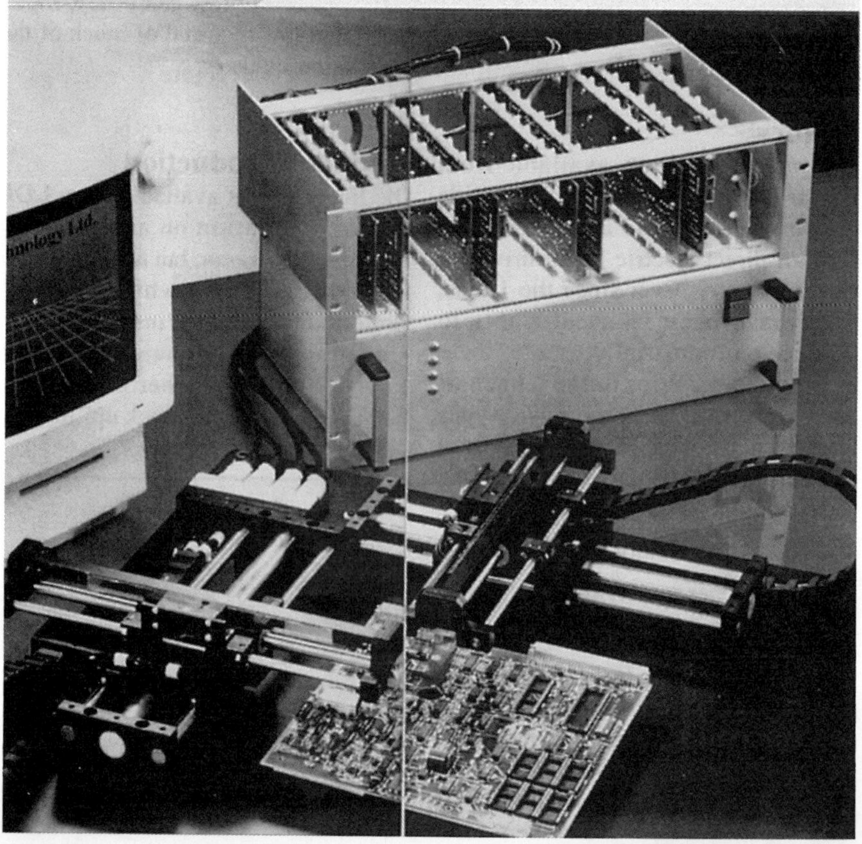

APPENDIX C

A comparison of the parameters of three different neodymium–iron–boron d.c. linear motors

Parameters	Motor 1	Motor 2	Motor 3
Overall length (cm)	8.0	12.0	23.7
Outside diameter (cm)	13.2	13.2	13.2
Total magnet area (cm^2)	13.5	27.0	54.0
Weight (kg)	4.5	5.4	8.3
Armature current (A)	50	50	50
Output force (N)	18.9 (at 0.9 m s^{-1})	17.6 (at 1.1 m s^{-1})	18 (at 1.1 m s^{-1})
Output power (W)	19.0 (at 2.3 m s^{-1})	21.0 (at 2.4 m s^{-1})	23.0 (at 2.5 m s^{-1})
Maximum power-to-weight ratio (W kg^{-1})	4.3	3.9	2.8
Maximum efficiency (%)	31	34	37

REFERENCES

1. Abel, E., Corbett, A. E., Mulhall, B. E., and Rhodes, R. G. (1974). *Levitation and propulsion of guided vehicles using superconducting magnets*. IEE Conference on Linear Electric Machines, IEE Publication No.120, pp. 223–9, IEE, London.
2. Anayi, F. J. (1993). Investigation of a brushless d.c. NdFeB magnet linear motor. Unpublished Ph.D. thesis. University of Wales.
3. Anayi, F. J. and Basak, A. (1989). *Brushless Nd–Fe–B magnet d.c. linear motors*. Fourth International Conference on Electrical Machines and Drives, pp. 71–75, IEE, London.
4. Armstrong, A. G., Riley, C. P., and Simkin, J. (1982). *TOSCA user guide*, Rutherford Appleton Laboratory, Oxford, UK,
5. Bahmanyar, H. and Ellison, A. J. (1974) . *The application of permanent magnets to the suspension of surface-guided vehicles*. IEE Conference on Linear Electric Machines, IEE Publication No. 120, pp. 203–9, IEE, London.
6. Bailey, A. R. and Edgar, J. P. (1974). *The use of linear motors in textile shuttle propulsion*. IEE Conference on Linear Electric Machines, IEE Publication No. 120, pp. 101–6, IEE, London.
7. Bamji, P. J. F. (1974). *Some applications of a linear electric machine for handling molten aluminium*. IEE Conference on Linear Electric Machines, IEE Publication No. 120, pp. 68–76, IEE, London.
8. Bannister, B. R. and Whitehead, D. G. (1986). *Transducers and interfacing*. Van Nostrand Reinhold, London.
9. Basak, A. and Al-Doori, T.H. (1984). *Microprocessor controlled brushless d.c. linear motor*. International Conference on Power Electronics and Variable-speed Drives, IEE, Savoy Place, London, pp. 402–4
10. Basak, A. and Flores-Filho, A. F. (1994). *Design, analysis and test of a novel slotless linear stepping motor*. Proceedings of International Conference on Electrical Machines, **2**, pp. 98–100, Paris.
11. Basak, A. (1975). The ferrite field closed end linear motor. *Journal of Physics E*. **8**, 716–17.
12. Basak, A. (1979). Generalised theory of ferrite-field d.c. linear motors. *International Journal of Electrical Engineering Education*. **16**, 43–9.
13. Basak, A. (1975). An investigation of d.c. linear motors. Unpublished PhD thesis. University of Wales.
14. Basak, A. (1977). Measurement of flux in a ceramic permanent magnet

of complex shape. *International Journal of Electrical Engineering Education*, **14**, 141–4.

15. Basak, A., Matsumura, A., Kaczmarek, J., and Lloyd, G. (1994). Computation of flux and force in an $SmCo_5$ d.c. linear motor with U-spring supports. *Journal of Magnetism and Magnetic Materials*, **133**, 583–6.
16. Basak, A. and Shirkoohi, G. H. (1990). Computation of magnetic field in d.c. brushless linear motors built with NdFeB magnets. *IEEE Trans. on Magnetics*. **26**, (2), 948–51.
17. Basak, A. and Overshott, K. J. (1974). *The ferrite field d.c. linear motor*, IEE Conference on Linear Electric Machines, IEE Publication No. 120, pp. 39–44, IEE, London.
18. Basak A. and Simonen, S. A. E. (1991). *A linear motor as a fader for audio and lighting controls*. Universities Power Engineering Conference, pp. 650–3, University of Bath, Bath.
19. Becker, J. J., Luborsky, F. E., and Luther Martin, D. L. (1968). Permanent magnet materials. *IEEE Trans. on Magnetics*, **4,** (2), 84–99.
20. Belikov, V. T., Ivlev, A. D., and Sitnichenko, V. M. (1983). Linear direct-current electric motor for the electric drives of industrial mechanisms. *Elek-trotekhnika*. **54**, (1), 56–8.
21. British patent No 12364 (1895). Jacquard, Weaver of Electric Shuttle Company.
22. Byerte, L. A. (1962). An electromagnetic chute for molten metal. *Elektrichestvo*, **5**, 74–7.
23. Campbell, P. (1994). *Permanent magnet materials and their applications*. Cambridge University Press.
24. Chirgwin, K. M. (1974). *Test results from the U.S. linear induction motor research vehicle program*. IEE Conference on Linear Electric Machines, IEE Publication No. 120, pp. 236–43.
25. Dascalescu, L., Morar, R., and Iuga, A. (1987). *Ferrite field d.c. linear motors for electrode-positioning systems of electrostatic separators*. Electric Energy Conference, Adelaide, pp. 473–5.
26. Desmond, D. J. (1945). The economic utilization of modern permanent magnets. *JEEE*. **92**, (2) 229–52
27. Doebelin, E. O. (1980). *System modelling and response: theoretical and experimental approaches*. Wiley, New York.
28. Ensminger, D. (1973). *Ultrasonic–the low and high intensity applications*. Dekker.
29. Fruechte, R. D., Nelson, R. H., and Radomski, T. A. (1977). *Power conditioning systems for a magnetically levitated test vehicle*. General Motors Research Laboratories. Report GMR-243.
30. Gerrard, J. and Paul, R. J. A. (1971). Rectilinear screw-thread reluctance motor. *Proc. IEE*, **118**, (11), 1575–84.
31. Green, C. W. and Paul, R. J. A. (1969). Application of d.c. linear machines

as short-stroke and static actuators. *Proc. IEE*, **116**, (4), 599–604.
32. Green, C. W. and Paul, R. J. A. (1969). *Factors involved in the analysis and design of DC linear actuators.* Proceedings of the IFAC Congress, Warsaw, pp. 116–37.
33. Griffiths, J. T. and Jones, P. L. (1969). *The direct current linear motor and its applications.* Universities Power Engineering Conference, University of Nottingham, pp. 89–91.
34. Gould, J. E. (1970). Some aspects of permanent magnet d.c. motors. *IEEE Trans. on Magnetics*, **6**, (2), 275–8.
35. Hadfield, D. (1962). *Permanent magnets and magnetism*, Iliffe, London.
36. Hanaoka, M., Miyaji, H., Kano, Y., Hasebe, S., and Ohta, K. (1992). Analysis of moving magnet d.c. linear motor with two straight stators. *Electrical Engineering in Japan.* **112**, (5), 100–7.
37. Hanrahan, D. J. and Toffolo, D. S. (1957). Permanent magnet generators: part 1–theory. *Trans. AIEE.* **76**, (3), 1098–1101.
38. Hanrahan, D. J. and Toffolo, D. S. (1963). Permanent magnet generators: part 2–optimum design. *Trans. AIEE.* **82**, 68–74.
39. Heidenhain (GB) Limited, *NC linear encoder.* Catalogue, obtainable from: Heidenhain (GB) Limited, 200 London Road, Burgess Hill, West Sussex, RH15 9RD.
40. Hochhausler, P. (1974). *A catamaran as a magnetically levitated vehicle.* IEE Conference on Linear Electric Machines, IEE Publication No. 120, pp. 216–22.
41. Honds, L. and Meyer, K. H. (1982). A linear d.c. motor with permanent magnets. *Philips Technical Review.* **40**, (11/12), 329–37.
42. Farrand Controls, *Linear Inductosyn® Scale.* Catalogue, obtainable from: Farrand Controls, 99 Wall Street, Valhalla, NY 10595, USA.
43. Jones, P. L. (1972). *Report No. ECRC/N536*, The Electricity Council Research Centre, (September)
44. Kamerbeck, E. M. H. (1973). Electric motor. *Philips Technical Review.* **33,** (8/9), 215–34.
45. Kano, Y., Matsumura, A., and Nagai, J. (1989). Moving magnet type linear d.c. motor by wireless control. Unpublished research report. University of Tokyo of Agriculture and Technology, Japan.
46. Kenjo, T. and Nagamori, S. (1985). *Permanent magnet and brushless d.c. motors.* Clarendon Press, Oxford.
47. Laithwaite, E. R. (1973). Linear motors for high speed vehicles. *New Scientist*, (28 June), 802–5.
48. Laithwaite, E. R. and Barwell, F. T. (1969). Application of linear induction motors to high-speed transport systems. *Proc. IEE*, **116**, 713–24.
49. Laithwaite, E. R. and Duxbury, J. (1957). Electromagnetic shuttle propelling devices. *Journal of the Textile Institute*, **48,** 211–24.
50. Laithwaite, E. R. and Lawrenson, P. J. (1957). A self-oscillating induc-

tion motor for shuttle propulsion. *Proc. IEE,* **104A,** (14), 93-–101.
51. Laithwaite, E. R. and Mamak, R. S. (1962). An oscillating synchronous linear machine. *Proc. IEE*, **109A**, 415–20.
52. Laithwaite, E. R. and Nix, G. F. (1960). Further developments of the self oscillating induction motor. *Proc. IEE*, **107A**, 476–86.
53. Lindemann, M. (1994). Whispers from the runway. *Financial Times*, (15 Dec.), p. 22.
54. Lindsley, J. C. (1971). Multipole closed end linear motor. *IBM Technical Disclosure Bulletin.* **13**, (12), pp. 3682–3.
55. Leung, S. W., Chow, T. S., and Zhu, Y. S. (1991). *An application of linear motor to loudspeaker systems.* Conference on Electrical Machines and Drives, IEE Publication No. 341, pp. 106–9.
56. Matsui, K., Umemori, T., Taketsuna, Y., and Hosada, Y. (1974). *DC linear motor controlled by thyristors and the testing equipment for its high_speed characteristics.* Proc. IEE Conference on Linear Electric Machines, IEE Publication No. 120, pp. 149–54.
57. Matsumura, A., Kano, Y., and Basak, A. (1993). A $SmCo_5$ field direct current linear motor with U-spring supports. *Journal of Applied Physics.* **73**, (10), 5632–4.
58. Matsumura, F., Maeda, S., Fujita, M., and Fukuzono, H. (1987). Completely contactless linear d.c. motor using magnetic suspension. *Electrical Engineering in Japan.* **107**, (1), 95–103.
59. McLean, G. W. (1988). Review of recent progress in linear motors. Proc. *IEE*, **135B**, (6), 380–416.
60. Messerle, H.G.(1965). *Dynamic circuit theory*. Pergamon Press, New York.
61. Midgley, G., Howe, D., and Mellor P. H. (1992). *The design and analysis of linear variable displacement transducers.* Universities Power Engineering Conference, pp.764–767.
62. Moczala, H. (1985). Burstenlose gleichstromlinearmotoren kleiner leistung. *Electrotech.* **106**, 1080–4
63. Pantic, B. O. (1974). *Linear oscillating electric machine–a synchronous condenser*, IEE Conference on Linear Electric Machines, IEE Publication No. 120, pp. 112–7.
64. Paul, R. J. A. (1972). Flat single-sided linear helical reluctance motor. *Proc. IEE*, **119**, (12), 1693–7.
65. Paul, R. J. A. and Reid, G. G. (1974). *Stepping-mode performance of a linear helical reluctance motor.* IEE Conference on Linear Electric Machines, IEE Publication No. 120, pp. 77–82.
66. Persson, E. K. (1983). *Brushless motor technology*. Motor-con Proceedings. pp. 129–40.
67. Ratcliff, G. and Griffiths, J. (1964). A linear d.c. motor. *Journal of Scientific Instruments*, **41**, 267–8.

68. Roters, H. C. (1970). *Electromagnatic devices*. Wiley, New York.
69. Ryashentev, N. P., Malov, A. T., Ugarov, G. G., and Fedonin, V. N. (1974). *Application of a general analysis for single-sided linear induction motors*. Proc. IEE Conference on Linear Electric Machines, IEE Publication No. 120, pp. 197–202.
70. Say, M. G. (1971). *Introduction to the unified theory of electromagnetic* machines. Pitman, London.
71. Schultz, R. A., (1984). Linear d.c. brushless motor. *IBM Technical Disclosure Bulletin*. **27**, (1A), pp. 50–4.
72. Shaw, J. M., Sketch, H. J. H., and Logie, J. M. (1954). *The theory, design, construction and testing of an electric launcher*. RAE Farnborough, Report No. AERO 2523, EL 1484.
73. The Math Works, Inc. *Simulink®*, Catalogue, obtainable from: The Math Works, Inc. 24 Prime Park Way, Natick, MA 01760-1520, USA.
74. Smith, A. C. and Appel, L.C. (1987). *Determining the maximum thrust of a permanent magnet linear actuator*. Third International Conference on Electrical Machines and Drives, London, pp. 344–7.
75. Thorn, K. and Norwood, J. (1961). *Theory of an electromagnetic mass accelerator for achieving hypervelocities*. NASA Technical Note: D886.
76. Triplett, B. L. (1986). *DC linear motors*. Motor-Con Proceedings, pp.104–8.
77. US Department of Transportation. (1967). Study. *Contract C 145-66, PB174866*.
78. Usami, Y., Fujie, J., and Fujiwara, S. (1974). *Studies on linear motor in the institute of JNRE*. Conference on Linear Electric Machines, IEE Publication No. 120, pp. 131–36.
79. Waite, M. (1994). *Multi-axis carrier for industrial applications*. Unpublished M.Sc. Dissertation, University of Wales.
80. Wakiwaka, H., Ninomiya, T., Morimura, T., and Yamada, H. (1994). Analysis of magnetic field and dynamic responses of linear d.c. motor for pen recorder.In *Simulation and design of applied electromagnetic systems*, (ed. T. Honma). Elsevier Science, Amsterdam, pp. 573–6.
81. West, J. C. and Jayawant, B. V. (1962). A new linear oscillating motor. *Proc. IEE*, **109A**, 292–300.
82. Wilson, H. W. (1905). *Transactions of the Liverpool Engineering Society*, **26**,
83. Woollen, W. B. (1955). Electromagnetic pumping of liquid metals. *Fluid Handling*, 90–2.
84. Yamada, H., Ota, H. Ogasawara, T., and Watabe, Y. (1987). Proposal of a blood circulation simulator which utilizes two linear d.c. motors. *IEEE Trans. on Magnetics*, **23**, (5), 2434-6.
85. Zehden, A. (1905), US Patent No. 732312.

INDEX